AS/A-LEVEL YEAR 1

STUDENT GUIDE

EDEXCEL

Physics

Topics 4 and 5

Materials
Waves and the particle nature of light

Mike Benn

PHILIP ALLAN FOR
HODDER
EDUCATION
AN HACHETTE UK COMPANY

Special thanks to Graham George for his help and advice in the publication of this book.

Philip Allan, an imprint of Hodder Education, an Hachette UK company, Blenheim Court, George Street, Banbury, Oxfordshire OX16 5BH

Orders

Bookpoint Ltd, 130 Park Drive, Milton Park, Abingdon, Oxfordshire OX14 4SE

tel: 01235 827827

fax: 01235 400401

e-mail: education@bookpoint.co.uk

Lines are open 9.00 a.m.–5.00 p.m. Monday to Saturday, with a 24-hour message answering service. You can also order through the Hodder Education website: www.hoddereducation.co.uk

© Mike Benn 2015

ISBN 978-1-4718-4359-4

First printed 2015

Impression number 5 4 3 2 1

Year 2019 2018 2017 2016 2015

This guide has been written specifically to support students preparing for the Edexcel AS and A-level physics examinations. The content has been neither approved nor endorsed by Edexcel and remains the sole responsibility of the author.

Cover photo: alexskopje/Fotolia

Typeset by Integra Software Services Pvt Ltd, Pondicherry, India

Printed in Italy

Hachette UK's policy is to use papers that are natural, renewable and recyclable products and made from wood grown in sustainable forests. The logging and manufacturing processes are expected to conform to the environmental regulations of the country of origin.

Contents

Getting the most from this book 4

About this book 5

Content Guidance

Materials 6

   Properties of fluids 6

   Tensile behaviour of solid materials 14

   Properties of solid materials 18

Waves and the particle nature of light 20

   Wave terminology 20

   Longitudinal and transverse waves 22

   Electromagnetic waves 23

   Superposition and interference 25

   Standing waves 31

   Refraction 37

   Lenses 41

   Plane polarisation 44

   Diffraction 46

   Reflection 49

   Particle nature of light 51

   Planck's equation 52

   Photoelectric effect 53

   Atomic spectra 56

   Wave–particle duality 58

Questions & Answers

AS Test Paper 61

A-level Test Paper 77

Knowledge check answers 94

Index 95

◼Getting the most from this book

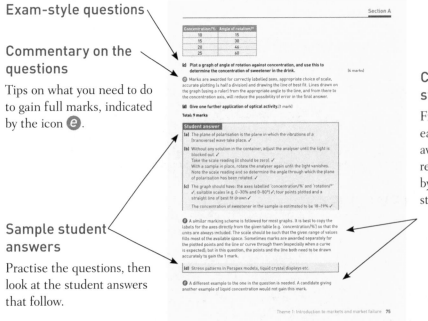

■About this book

This guide is one of a series covering the Edexcel specification for AS and A-level physics. It offers advice for the effective study of Topic 4 (Materials) and Topic 5 (Waves and the particle nature of light) of the specification. Its aim is to help you to *understand* the physics — it is not intended as a shopping list that enables you to cram for the examination. The guide has two sections.

■ The **Content Guidance** is not intended to be a detailed textbook. It offers guidance on the main areas of the content of the topics, with an emphasis on worked examples. These examples illustrate the types of question you are likely to come across in the examinations.

■ The **Questions & Answers** section comprises two sample tests — one has the same structure and style as the AS paper 2 examination and the other has the structure of an A-level paper 2. Both tests are restricted to the content of this guide. Answers are provided and, in some cases, distinction is made between responses that might have been given by an A-grade candidate and those of a typical C-grade candidate. Common errors made by candidates are also highlighted so that you, hopefully, do not make the same mistakes.

The purpose of this book is to help you with the AS paper 2, A-level paper 2 and synoptic paper 3 examinations — but don't forget that what you are doing is learning physics. The development of an understanding of physics can only evolve with experience, which means time spent thinking about physics, working with it and solving problems. This book provides you with a platform to do this.

If you try all the worked examples and the tests before looking at the answers, you will begin to think for yourself, as well as develop the necessary techniques for answering examination questions. In addition, you will need to learn the basic formulae, definitions and experiments. Thus prepared, you will be able to approach the examination with confidence.

The specification states the physics that will be examined in the AS and A-level examinations and describes the format of those tests. This is not necessarily the same as what teachers might choose to teach (or what you might choose to learn).

The specification can be obtained from Edexcel, either as a printed document or from the web at www.edexcel.com.

Content Guidance

■ Materials

Properties of fluids

A fluid is a material that flows. Unlike a solid, in which the atoms occupy fixed positions, the particles of a fluid can move relative to each other. For this topic fluids will be considered as liquids and gases only, but you should be aware that plasma in stars and some solids such as glass also exhibit fluid characteristics.

Density of fluids

Density is a property used to compare the masses of equal volumes of different materials.

$$\text{density} = \frac{\text{mass}}{\text{volume}} \quad \rho = \frac{m}{V} \quad \text{unit: kg m}^{-3}$$

Density is mass per unit volume.

Fluid	Density/kg m^{-3}	Fluid	Density/kg m^{-3}
Mercury	13 600	Carbon dioxide	1.78
Sulfuric acid in a car battery	1250	Air	1.24
Water	1000	Helium	0.161
Ethanol	790	Hydrogen	0.081

Table 1

Table 1 shows the wide range of density of fluids. Gases have much lower densities than liquids because the molecular separation is much larger. Liquids are virtually incompressible, but the density of gases will increase with increasing pressure. The values in the table are for pressures of 1.01×10^5 Pa at 293 K. The relationships between pressure, temperature and volume of gases will be studied in more detail in year 2 of the course.

Worked example

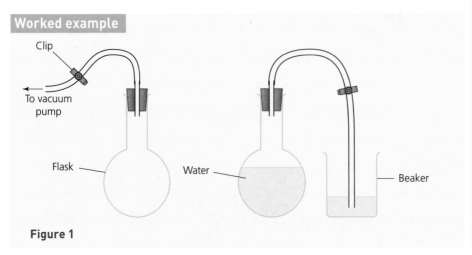

Figure 1

In an experiment to measure the density of air, the mass of a glass flask is measured before and after removing most of the air using a vacuum pump (Figure 1). The volume of the air removed at normal pressure is found by putting the end of the tubing into a beaker of water and releasing the clamp.

The measurements are given below:
- mass of flask plus air = 212.32 g
- mass of flask after pumping = 211.35 g
- volume of water drawn into flask = 785 cm^3

Use these measurements to determine the density of air.

Answer

mass of air removed = 212.32 g − 211.35 g = 0.97 g = 9.7×10^{-4} kg

volume of air removed = 785 cm^3 = 7.85×10^{-4} m^3

$$\text{density} = \frac{9.7 \times 10^{-4}\text{ kg}}{7.85 \times 10^{-4}\text{ m}^3} = 1.24 \text{ kg m}^{-3}$$

Exam tip

Determination of density simply requires the measurement of mass and volume. A balance and measuring cylinder can be used for liquids.

Pressure in fluids

At any point in a column of fluid there is a pressure that acts equally in all directions, the value of which depends on the height of the fluid above that point (Figure 2).

$$\text{pressure} = \frac{\text{force}}{\text{area}} = \frac{\text{weight of column}}{\text{area}} = \frac{mg}{A} = \frac{V\rho g}{A} = \frac{(Ah)\rho g}{A} = h\rho g$$

Worked example

Estimate the height of the Earth's atmosphere given that the density and the air pressure at the surface are 1.24 kg m^{-3} and 1.01×10^5 Pa respectively.

Answer

Assuming the average density to be about 0.6 kg m^{-3} and taking g = 10 m s^{-2}:

$$1 \times 10^5 \text{ Pa} = h \times 0.6 \text{ kg m}^{-3} \times 10 \text{ m s}^{-2}$$

$$h = \frac{1 \times 10^5 \text{ Pa}}{0.6 \text{ kg m}^{-3} \times 10 \text{ m s}^{-2}} = 17 \text{ km} \approx 20 \text{ km}$$

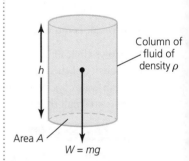

Column of fluid of density ρ

Area A

$W = mg$

Figure 2

It is important to be aware of the effect of the density of the fluid in pressure calculations. At one time, atmospheric pressure was measured in terms of the height of the column of mercury that balanced the air pressure using a mercury barometer. The standard value was 760 mm of mercury. A similar pressure will be exerted by water at a depth of 10 m. This means that a diver will experience a total pressure of about 2 atmospheres at this depth (1 atm due to air pressure and 1 atm due to the water).

Knowledge check 1

Use the data in the table of densities to show that 10 m of water and 760 mm of mercury both exert a pressure of about 1×10^5 Pa.

Upthrust

If you are in a swimming pool, you will experience a buoyancy force that enables you to float or swim. This force is called an *upthrust* — it is a consequence of the water pressure being larger below an immersed object than above it (Figure 3).

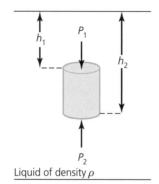

Liquid of density ρ

Figure 3

- pressure on the upper surface, $P_1 = h_1 \rho g$
- pressure on the lower surface, $P_2 = h_2 \rho g$
- $\Delta P = (h_2 - h_1)\rho g$
- upthrust, $\Delta F = \Delta P \times A = (h_2 - h_1)\rho g \times A = (h_2 - h_1)A\rho g = V\rho g$
- $\Delta F = mg$ = weight of the fluid displaced

This result is often stated as **Archimedes' principle**.

It follows that, for a given volume of displaced fluid, the upthrust is proportional to the density of the fluid. You will experience larger flotation forces in sea water than in fresh water because brine has a larger density. Also, the upthrust of the air on your body is so small that it is not noticeable because the density of air is very small.

Archimedes' principle states that the upthrust on a body immersed in a fluid is equal to the weight of the fluid displaced.

Knowledge check 2

What condition is required for an object to float in a liquid?

Exam tip

In many upthrust calculations you will be given the mass of the fluid displaced. Make sure that this is converted to weight by multiplying by $9.8\,\mathrm{m\,s^{-2}}$.

Worked example

An uninflated toy balloon has a mass of 4.9 g. It is filled with helium to form a sphere of diameter 24 cm, and is held by a length of string tied to a post so that it floats directly above the point of attachment (Figure 4). Use the data in the fluid density table (Table 1) to show that the tension in the string is about 0.03 N.

Answer

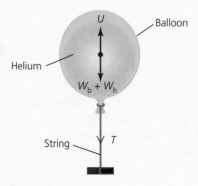

Figure 4

For equilibrium:

$U = W + T$

U = weight of displaced air

$$= \frac{4}{3}\pi \times (0.12 \text{ m})^3 \times 1.24 \text{ kg m}^{-3} \times 9.8 \text{ m s}^{-2} = 0.088 \text{ N}$$

W = weight of rubber + weight of helium = $W_r + W_h$

$$= (4.9 \times 10^{-3} \text{ kg} \times 9.8 \text{ m s}^{-2}) + \left(\frac{4}{3}\pi \times (0.12 \text{ m})^3 \times 0.161 \text{ kg m}^{-3} \times 9.8 \text{ m s}^{-2}\right)$$

$$= 0.059 \text{ N}$$

$T = 0.088 \text{ N} - 0.059 \text{ N}$

$$= 0.029 \text{ N} \approx 0.03 \text{ N}$$

Moving fluids

In fluid flow, particles in the fluid are affected by *cohesive* forces of neighbouring molecules and *adhesive* forces to the surfaces of obstructing solids or the inner walls of a pipe. The path taken by an individual particle in a moving fluid is called a **streamline**.

When the streamlines of adjacent particles do not cross over each other, the flow is said to be **laminar** (often called **streamlined flow**). When the paths do cross, the flow becomes **turbulent** (Figure 5).

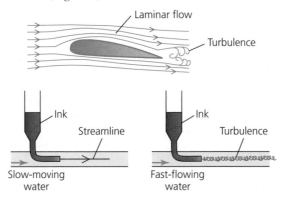

Figure 5 Laminar and turbulent flow

Figure 5 shows that for an aircraft wing the airflow needs to be laminar around the wing, although some turbulence is likely beyond the wing. For fluids in a tube the flow will be streamlined at low rates, becoming turbulent at a critical velocity when the fluid swirls around forming **vortices** or **eddy currents**.

Viscosity

When a fluid moves through a pipe, or flows relative to a solid body, it will experience resistive forces that impede its motion. The rate of flow depends on the **viscosity** (the degree of 'stickiness') of the fluid. For example treacle and most oils will flow much less readily than water, and gases generally will have a much lower viscosity than liquids.

Exam tip

When drawing streamlines, make sure that the lines are continuous and never cross.

Relative values of the viscosities of liquids can be measured using a simple **viscometer** as shown in Figure 6. The time taken for the liquid level in the reservoir to fall between the two marks is measured for a range of different liquids.

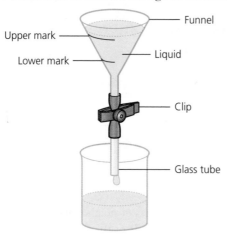

Figure 6 A simple viscometer

The rate of flow of a fluid though a tube depends on:

- the viscosity of the fluid
- the diameter of the tube
- the length of the tube
- the pressure across its ends
- whether the flow is streamlined or turbulent

The most significant factor is the diameter of the tube. If the diameter is doubled then the flow rate increases by sixteen times if all other factors remain the same. The diameters of oil pipes and gas pipes need to be as large as possible to enable an economic flow rate. Restriction in blood flow in narrowed arteries is a common cause of high blood pressure.

Stokes' law

Stokes' law relates to the special case of a spherical body falling through a fluid in which the flow of the fluid relative to the body is streamlined.

- viscous drag, $F = 6\pi\eta rv$
- η is the coefficient of viscosity (N s m^{-2})
- r is the radius of the sphere
- v is the relative velocity of the fluid to the sphere

In addition to the motion of solid spheres in liquids, Stokes' law is used in the study of raindrops, mist particles and aerosol droplets in the atmosphere.

The free-body diagram for a sphere falling through a fluid (Figure 7) shows that the resultant force acting on the sphere will be given by:

$$F = W - (U + D)$$

Initially, when the sphere is held stationary, the viscous drag is zero and the resultant downward force equals the weight minus the upthrust. On release, the sphere accelerates downwards and, as the velocity increases, so does the viscous drag, D. The resultant force is therefore reduced until it becomes zero, when the sphere is in equilibrium and it stops accelerating. The sphere will then continue to fall with a constant velocity known as its **terminal velocity**.

Measuring the terminal velocity of a sphere falling through a fluid provides a method for obtaining a value for the viscosity of the fluid.

When the terminal velocity is reached:

$W - (U + D) = 0$ so, $W = U + D$

$$\frac{4}{3}\pi r^3 \rho_s g = \frac{4}{3}\pi r^3 \rho_f g + 6\pi\eta rv$$

(weight = mg = volume of sphere × density × g; ρ_s and ρ_f are the densities of the solid and the fluid)

$$\eta = \frac{2(\rho_s - \rho_f)gr^2}{9v}$$

Knowledge check 4

State the factors that will affect the terminal velocity of a sphere falling through a fluid. Explain which one has the most significant effect on the magnitude of the terminal velocity.

Core practical 4

Using a falling-ball method to determine the viscosity of a liquid

Core practical 4 requires you to determine the viscosity of a liquid using a falling-ball method. Full details of how the measurements are made, what precautions are taken and how the viscosity can be worked out from a graph of terminal velocity against r^2 may be needed for the examination.

Worked example

In an experiment to determine the viscosity of some motor oil at room temperature, a steel ball-bearing is released just below the surface of the oil and timed as it crosses a series of equally spaced marks on the glass cylinder (Figure 8). The experiment is repeated three times, and the average time, Δt, taken for the ball to cross each 5 cm division is calculated.

The results of the experiment were:

- separation of marks = 5.0 cm
- diameter of ball bearings = 3.0 mm
- density of steel = 7800 kg m^{-3}
- density of oil = 820 kg m^{-3}
- a Copy and complete Table 2 to include the incremental times and the average velocity between successive marks.

Figure 8

Mark	Average t/s	Δt/s	$v \times 10^{-2}$/m s^{-1}
1	1.33		
2	2.42		
3	3.40		
4	4.31		
5	5.22		
6	6.13		

Table 2

b Sketch a graph to show how the velocity of the ball changes as it falls through the oil.

c Determine the value of the critical velocity, and hence calculate the viscosity of the oil.

Answer

a

Mark	Average t/s	Δt/s	v/10^{-2}m s^{-1}
1	1.33	1.33	3.80
2	2.42	1.09	4.59
3	3.40	0.98	5.10
4	4.31	0.91	5.49
5	5.22	0.91	5.49
6	6.13	0.91	5.49

b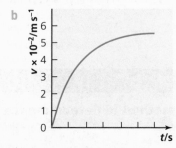

Figure 9

c Terminal velocity $v = 5.5 \times 10^{-2}\,\text{m s}^{-1}$

$$\eta = \frac{2(\rho_s - \rho_f)gr^2}{9v}$$

$$= \frac{2 \times (7800 - 820)\,\text{kg m}^{-3} \times 9.8\,\text{m s}^{-2} \times (1.5 \times 10^{-3}\,\text{m})^2}{9 \times 5.5 \times 10^{-2}\,\text{m s}^{-1}}$$

$$= 0.62\,\text{N s m}^{-2}$$

Exam tip

Remember that the gradient of a velocity–time graph represents the acceleration of the falling body. As the body falls, its acceleration, and hence the gradient, decreases until the line is horizontal. At this stage the acceleration is zero, and so the terminal velocity has been reached.

Viscosity of gases

The coefficient of viscosity of gases is much lower than that of liquids so the effect is less apparent, unless the flow rates are high. For domestic gas boilers, close to the supply, a short length of 15 mm diameter copper pipe is sufficient for an adequate flow, but for appliances requiring several metres of tubing with some bends, the regulations stipulate the use of 22 mm tubing.

Because of the low viscosity of air (about $2 \times 10^{-5}\,N\,s\,m^{-2}$), the terminal velocity of falling objects is generally very large, but it depends a great deal on the size of the body. For bodies falling through air we can usually ignore the upthrust, because the air density is so small, so an object falling from rest will initially experience no upward forces and will begin to accelerate downwards at $9.8\,m\,s^{-2}$. As the speed increases the viscous drag also increases. A velocity–time graph similar to the one in the above example will represent this motion. The initial gradient will be g and the terminal velocity is the value when the curve has levelled off.

Exam tip

For solids and liquids falling through air, the upthrust (weight of air displaced) is very small compared to the other forces, and so it can be omitted from the calculations.

Worked example

Estimate the terminal velocity of a raindrop of diameter 1 mm and a droplet of mist of diameter 0.1 mm (viscosity of air = $2 \times 10^{-5}\,N\,s\,m^{-2}$, density of water = $1000\,kg\,m^{-3}$).

Answer

Ignoring the upthrust, at equilibrium:

weight of drop = viscous drag

$$\frac{4}{3}\pi r^3 \rho g = 6\pi \eta r v$$

which rearranges to

$$v = \frac{2r^2 \rho g}{9\eta}$$

For a 1 mm diameter:

$$v = \frac{2 \times (0.5 \times 10^{-3}\ m)^2 \times 1000\ kg\ m^{-3} \times 10\ m\ s^{-2}}{9 \times 2 \times 10^{-5}\ N\ s\ m^{-2}}$$

$$= 30\ m\ s^{-1}$$

For a 0.1 mm diameter:

$$v = \frac{2 \times (0.05 \times 10^{-3}\ m)^2 \times 1000\ kg\ m^{-3} \times 10\ m\ s^{-2}}{9 \times 2 \times 10^{-5}\ N\ s\ m^{-2}}$$

$$= 0.3\ m\ s^{-1}$$

In nature, the larger drops actually fall at about $5\,m\,s^{-1}$, but the speed of the smaller drops is about right. It follows that Stokes' law applies to the slower small droplets but not to the larger ones.

Summary

After studying this topic, you should be able to:

- define density and be aware of the factors affecting the densities of liquids and gases
- explain how fluid pressure depends on the fluid's density and its depth, and use the equation $P = h\rho g$ to calculate or compare fluid pressures
- understand the difference between laminar flow and turbulent flow
- appreciate the meaning of the viscosity of a fluid, and use Stokes' law to calculate viscous forces
- understand how objects falling through a fluid can reach terminal velocity; use free-body force diagrams to determine this velocity

Tensile behaviour of solid materials

'Tensile' involves linear extensions or compressions due to applied forces. The behaviour of many materials subjected to tensile forces is similar to that of a spring.

Hooke's law

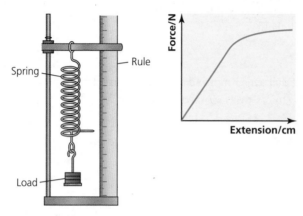

Figure 10 Demonstrating Hooke's law

A graph of force against extension shows that the extension is directly proportional to the load until a spring is overloaded and begins to lose its elasticity (Figure 10). When in the region of proportionality, the spring obeys **Hooke's law**.

- $F = k\Delta x$
- k is the **spring constant** and is equal to $\dfrac{F}{\Delta x}\ \mathrm{N\,m^{-1}}$, which is the gradient of the linear region of the graph.

You will see later that most metals obey the law for relatively low loads, but many polymers do not, whatever the load.

> **Hooke's law** states that, up to a certain limit, the extension is directly proportional to the load.

Force–extension graphs

The properties of solid materials are readily illustrated using force–extension graphs. An experiment in which a long piece of copper wire is loaded until it breaks, taking corresponding values of load and extension, produces a force–extension graph (Figure 11) that is typical of most metals.

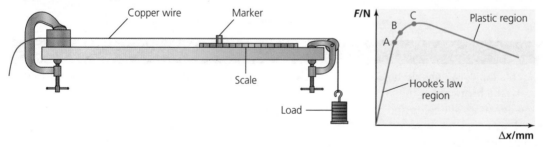

A – limit of proportionality
B – elastic limit
C – yield point

Figure 11

The following features need to be known and explained in the examination:

- **Hooke's law region** — where F is proportional to Δx.
- **limit of proportionality** — the point where Hooke's law ceases to be obeyed.
- **elastic limit** — the point beyond which the wire will not regain its original length when the load is removed.
- **plastic flow** — the region where the metal deforms plastically. If the load is removed there will be little or no change in the extended length of the wire.
- **yield point** — the onset of plastic flow.

When a metal behaves *elastically* the bonds holding the atoms together are stretched. When the stretching force is removed the atoms return to their original positions and the metal regains its shape.

At the yield point, layers of atoms slide over one another and cannot return to their initial positions. This is called 'plastic flow'.

Data for drawing force–compression graphs are more difficult to obtain, unless specialised equipment, such as a hydraulic press, is available. The elastic region is much the same as in extension graphs, but after the Hooke's law region the metal rods tend to buckle and break.

Polymers such as plastics and rubber have more complex molecular structures than metals and so behave differently when deformed. If the spring in Figure 11 was replaced by a rubber band, a force–extension graph like that shown in Figure 12 would be result.

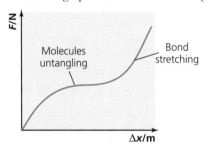

Figure 12 Force–extension curve for a rubber band

Polymers have long-chain molecules that are initially in a 'tangled' state, a like a pile of cooked spaghetti. When a tensile force is applied, the cross-links between the polymer molecules are broken and the chains straighten out, leading to a large extension for a relatively small additional load. When the polymer molecules become untangled, the load has to stretch the strong chemical bonds and so the rubber becomes much more difficult to stretch.

Stress, strain and the Young modulus

Force–extension graphs give information relating to a particular specimen of a solid material, but these curves vary for different-sized samples of the same material. In order to make the graphs consistent for a given material, the dimensions must be taken into account. In order to do this, the quantities **stress**, **strain** and **Young modulus** are defined for materials that are subjected to tensile forces.

Exam tip

You will be required to describe the behaviour of a material with reference to the graph in Figure 11, and may have to label the appropriate points in the examination.

Exam tip

In examinations, most graphs will be drawn as shown in this section. However it is general practice to plot the independent variable (the applied force) on the x-axis, so that experimental results can be represented on an extension–force graph.

stress $\sigma = \dfrac{F}{A}$ unit: pascal, Pa (N m^{-2})

strain $\varepsilon = \dfrac{\Delta l}{l}$ no unit

Young modulus $E = \dfrac{\sigma}{\varepsilon}$

unit: Pa

Core practical 5

Determine the Young modulus of a material

Core practical 5 requires you to determine the Young modulus of a material by adding weights to a long piece of wire. Full details of the measurements made, the precautions to be taken and how the Young modulus is calculated from a stress–strain graph may be needed in the examination.

Knowledge check 5

State the measurements you would need to make to determine the Young modulus of a length of wire. Suggest an appropriate instrument for each quantity you would measure.

Worked example

A steel wire of length 2.00 m and diameter 0.40 mm is extended by 4.0 mm when a tensile force of 50 N is applied. Calculate:

a the stress

b the strain

c the Young modulus of steel

Answer

a Stress, $\sigma = \dfrac{50 \text{ N}}{\pi(0.20 \times 10^{-3}\,\text{m})^2} = 4.0 \times 10^8 \text{ Pa}$

b Strain, $\varepsilon = \dfrac{0.40 \times 10^{-3} \text{ m}}{2.00 \text{ m}} = 2.0 \times 10^{-4} \; (= 0.02\%)$

c Young modulus, $E = \dfrac{4.0 \times 10^8 \text{ Pa}}{2 \times 10^{-4}} = 2.0 \times 10^{12} \text{ Pa}$

Exam tip

You should be aware that solids under compression behave in a similar fashion to that experienced when tensile forces are applied — but beyond the elastic limit they will buckle rather than extend plastically.

Stress–strain graphs

The shape of the stress–strain graphs for metals and polymers is much the same as the force–extension graphs. For relatively small extensions of a long wire, the change in cross-sectional area is very small, so the stress is approximately proportional to the load, and the strain is always proportional to the extension.

Stress–strain graphs have the advantage of being applicable to the material under test, and not just a particular sample. In the Hooke's law region, where stress is proportional to strain, the gradient of the line equals the Young modulus of the material.

The ultimate tensile stress (UTS), or breaking stress, is the maximum stress a material can bear before it breaks. It is a surprising fact that the silk of a spider's web has a higher UTS than steel!

Worked example

The results of an experiment in which a length of copper wire is loaded until it breaks are given below:

Length of wire = 3.00 m   Mean diameter of wire = 0.52 mm

a Complete Table 3 to include the values of stress and strain.

Force/N	Extension/mm	Stress/Pa	Strain
0	0		
5.0	0.5		
10.0	1.0		
15.0	1.5		
20.0	2.0		
25.0	2.5		
30.0	3.0		
35.0	4.0		
40.0	6.0		
35.0	30.0		

Table 3

b Plot a stress–strain graph for the copper wire

c Use your graph to determine the Young modulus and then the UTS of copper.

Answer

a

Force/N	Extension/mm	Stress × 10^7/Pa	Strain × 10^{-4}
0	0	0	0
5.0	0.5	2.4	1.7
10.0	1.0	4.7	3.3
15.0	1.5	7.1	5.0
20.0	2.0	9.4	6.7
25.0	2.5	11.8	8.3
30.0	3.0	14.1	10
35.0	4.0	16.5	13
40.0	6.0	18.8	20
35.0	30.0	16.5	100

b

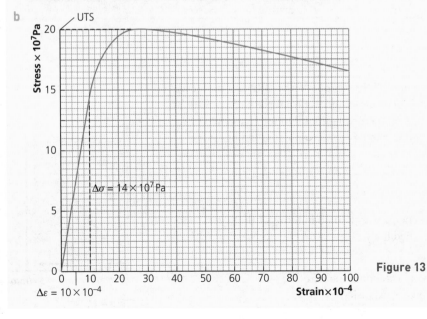

Figure 13

c The Young modulus is the gradient of the linear region:

$$\text{gradient} = \frac{14 \times 10^7 \text{ Pa}}{10 \times 10^{-4}} = 1.4 \times 10^{11} \text{ Pa}$$

The UTS is 2.0×10^8 Pa

Knowledge check 6

What are the advantages of stress–strain graphs over force–extension graphs in the study of the properties of a wire?

Properties of solid materials

So far you have explored the tensile behaviour of solid materials. You also need to be able to explain the meanings of a range of properties and describe how these are used in a variety of applications. Table 4 lists the properties, together with their descriptions or definitions, and also the name of the opposite property and some examples of materials that exhibit the property.

Property	Definition	Example	Opposite	Definition	Example
Strong	High breaking stress	Steel	Weak	Low breaking stress	Expanded polystyrene
Stiff	Gradient of a force–extension graph	Steel	Flexible	Low Young modulus	Natural rubber
Tough	High energy density up to fracturing; metal that has a large plastic region	Mild steel, copper, rubber tyres	Brittle	Little or no plastic deformation before fracture	Glass, ceramics
Elastic	Regains the original dimensions when the deforming force is removed	Steel in the Hooke's law region; rubber	Plastic	Extends extensively and irreversibly for a small increase in stress beyond the yield point	Copper, plasticine
Hard	Difficult to indent the surface	Diamond	Soft	Surface easily indented/scratched	Foam rubber, balsa wood
Ductile	Can be drawn into wires readily	Copper	Hard, brittle	(see above)	(see above)
Malleable	Can be hammered into thin sheets	Gold	Hard, brittle	(see above)	(see above)

Table 4

Exam tip

You should be able to compare some of the properties of different materials using force–extension graphs — for example brittle materials show little or no plastic flow, stiffness depends on the gradient and tough materials have extended plastic regions. Strengths can be compared using the breaking points on stress–strain graphs.

Elastic strain energy

In the energy section, elastic strain energy was defined in terms of the potential, or stored, energy in an elastically deformed body.

For a tensile extension:

elastic strain energy = work done in extending the wire
= average force × extension

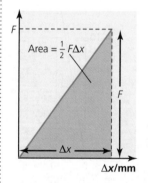

Figure 14

If Hooke's law is obeyed:

- $E_{el} = \frac{1}{2}F\Delta x$ = area under the line on the force–extension graph (Figure 14)
- $E_{el} = \frac{1}{2}(k\Delta x) \times \Delta x = \frac{1}{2}k(\Delta x)^2$

Even if Hooke's law is not followed it can be shown that the area under the line of any force–extension graph represents the elastic strain energy in the material.

For stress–strain graphs the area under the line represents the elastic strain energy per unit volume, commonly called the *energy density* of the deformed material.

Hysteresis in rubber

If a rubber band is loaded and then unloaded (by adding and then removing a number of equal masses) and the extensions are measured at each increment, the resulting force–extension graph will be as shown in Figure 15.

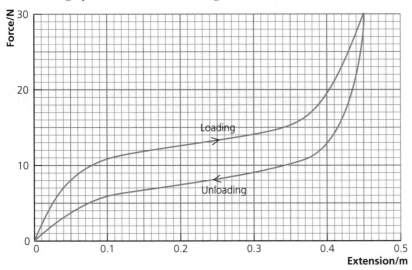

Figure 15 Loading and unloading curves for a rubber band

The graph shows that the unloading curve lies beneath the loading curve. This is known as **hysteresis** and can be explained in terms of elastic strain energy.

- During loading, work is done on the rubber. This is stored as elastic strain energy and is represented by the area under the loading curve.
- During unloading, the energy stored in the rubber does work by raising the load. This is represented by the area under the unloading curve.
- More energy has been transferred in stretching the band than has been transferred during unloading. By the law of conservation of energy this 'lost' energy must be transferred to another form. This will be in the form of internal energy within the rubber, which will increase the temperature of the band and then be dispersed as thermal energy to the surroundings. For example, each time a car tyre rotates, the rubber goes through a hysteresis cycle and so, after a long journey, the tyres will get warm.
- The increase in internal energy within the rubber band during one loading–unloading cycle is equal to the area enclosed by the loop.

Knowledge check 7

Car tyres are said to be 'tough'. How does the stress–strain graph for rubber help to explain this?

Exam tip

Always use arrows and clearly label the lines to distinguish the loading and unloading curves.

Worked example

Use the hysteresis graph for rubber (Figure 15) to estimate the increase in internal energy during the cycle shown.

Answer

The number of squares enclosed in the loop is about 8.

The area of each square represents $5\,\text{N} \times 0.05\,\text{m} = 0.25\,\text{J}$

So the internal energy is $8 \times 0.25\,\text{J} = 2\,\text{J}$

Exam tip

Other methods of estimating the area of a region in a graph — such as considering each curve as two triangles and a trapezium, and subtracting the lower area from the upper one — are usually quicker and more suitable in an examination situation.

Summary

After studying this topic, you should be able to:
- state Hooke's law and determine stiffness from a force–extension graph
- describe the behaviour of a metal wire when a tensile force is applied to its ends, and relate this to a force–extension graph
- determine the Young modulus of a material using force–extension and stress–strain graphs
- determine the elastic strain energy from a force–extension graph, and the energy density from a stress–strain graph
- describe the molecular behaviour of polymers when they are stretched and reformed, and explain what is meant by hysteresis

■ Waves and the particle nature of light

There are many types of wave. In this section you will study the properties of mechanical waves (such as sound waves) and electromagnetic waves (with particular emphasis on the behaviour of light). Most types of waves transfer energy from the source to the observer (they are progressive waves). However, in some cases (such as standing waves) energy is not transferred but is transformed between potential energy and kinetic energy.

Wave terminology

The oscillating particles in a mechanical wave, such as a stretched wire or slinky, can be described by a displacement–time graph for a single particle within the wave, or by the position of all the particles at a single instant along a section of the wave (displacement–distance graph — see Figure 16).

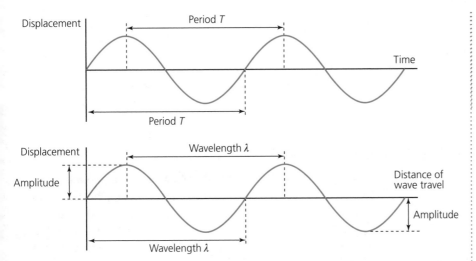

Figure 16

You need to learn the fundamental definitions that relate to wave motion. They are amplitude, period, frequency, wave speed and wavelength. It is useful to refer to the graphs in Figure 16 to gain a full understanding of their meaning.

- **Amplitude**, y_0 — the maximum displacement of a particle from the midpoint of the oscillation — unit: metre (m)
- **Period**, T — the time taken for one complete oscillation — unit: second (s)
- **Frequency**, f — the number of oscillations per second — unit: hertz (Hz)
- **Wave speed**, v — the distance travelled by the wave each second — unit: metres per second ($\mathrm{m\,s^{-1}}$)
- **Wavelength**, λ — the distance between consecutive points at which the oscillations are in phase — unit: metre (m)

From these definitions, two useful equations can be established:

i $\quad f = \dfrac{1}{T}$

    e.g. if there are 10 oscillations in 1 second, each oscillation will have a period of 0.1 s.

ii $\quad v = f\lambda$

    This is often called **the wave equation**, and comes from the fact that a wave travels one wavelength in the time taken for one oscillation, so that:

$$v = \frac{\lambda}{T} = \frac{1}{T} \times \lambda = f\lambda$$

> **Exam tip**
>
> It is worth noting that one complete cycle in a displacement–time graph represents one period, and one cycle of a displacement–distance graph covers a distance of one wavelength.

Worked example

The prongs of a tuning fork vibrate with a period of 2.5 ms. The speed of sound in air is $340\,\mathrm{m\,s^{-1}}$. Calculate:

a the frequency of the emitted sound

b the wavelength of the sound in air

$\rightarrow$

Answer

a $f = \dfrac{1}{T} = \dfrac{1}{2.5 \times 10^{-3}\ \text{s}} = 400\ \text{Hz}$

b $\lambda = \dfrac{v}{f} = \dfrac{340\ \text{m s}^{-1}}{400\ \text{Hz}} = 0.85\ \text{m}$

Longitudinal and transverse waves

The difference between **longitudinal** and **transverse** waves can be illustrated using a slinky spring (Figure 17).

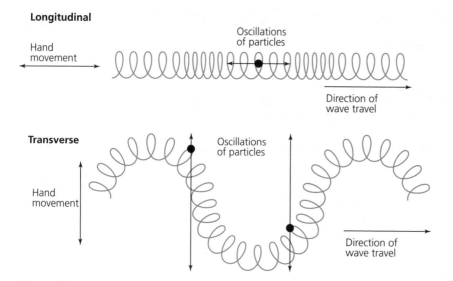

Longitudinal

Hand movement

Oscillations of particles

Direction of wave travel

Transverse

Hand movement

Oscillations of particles

Direction of wave travel

Figure 17

> In a **longitudinal** wave the particles oscillate backwards and forwards along the line along which the wave progresses.
>
> In a **transverse** wave the particles oscillate at right angles to the direction of propagation of the wave.

Sound waves are examples of longitudinal waves — the backwards and forwards motion of air molecules leads to alternate regions of high pressure and low pressure, called **compressions** and **rarefactions** respectively. The oscillations of the particles can be observed by putting a lighted candle close to a large loudspeaker emitting a low-frequency sound (Figure 18).

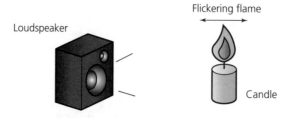

Flickering flame

Loudspeaker

Candle

Figure 18

If a stone is thrown into a pond, a series of circular ripples can be seen moving outwards from the point where the stone entered the water. A duck floating on the water will 'bob' up and down in the same place as the wave passes along the pond surface. Figure 19 shows the transverse wave motion of the water particles.

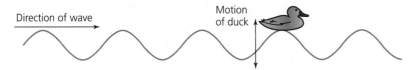

Figure 19

Other examples of transverse waves that you will encounter in this topic are waves in stretched strings or wires, and the variations of electric and magnetic fields in electromagnetic radiation.

Core practical 6

Determine the speed of sound in air

Core practical 6 requires you to determine the speed of sound in air using a double beam oscilloscope, signal generator, speaker and microphone. Full details of how the measurements are made, the precautions to be taken and how the speed of sound can be worked out from a graph of wavelength against the reciprocal frequency may be required for examinations.

Electromagnetic waves

Electromagnetic waves differ from the mechanical waves you have met so far — they do not consist of vibrating particles. These waves do not require a medium through which to travel because they consist of regularly changing electric fields and magnetic fields. You do not need to know the properties of these fields for the AS examination (these will be covered in Unit 4 of the A2 course), but you will need a general knowledge of the types of radiation that make up the electromagnetic spectrum, including similarities and differences between them, and be able to describe some of their applications.

All electromagnetic radiation:
- travels in a vacuum with a speed of $3.0 \times 10^8 \, \text{m s}^{-1}$
- consists of oscillating electric fields and magnetic fields that are in phase and whose transverse variations lie within planes at right angles to each other

Table 5 gives some information on the production, properties and wavelengths of the radiation types that make up the electromagnetic spectrum.

> **Exam tip**
>
> Electromagnetic waves are effectively transverse waves — but be aware that there are no particles vibrating.

Type of wave	Wavelength range/m	Method of production	Properties and applications
γ-rays	10^{-16} to 10^{-11}	Excited nuclei falling to lower energy states	Highly penetrating rays. Used in medicine for destroying tumours; diagnostic imaging; and sterilisation of instruments.
X-rays	10^{-14} to 10^{-10}	Fast electrons decelerating after striking a target	Similar to γ-rays, but the method of production means that their energy is more controllable. Used in medicine for diagnosis and therapy; in industry for detecting faults in metals; and studying crystal structures.

Type of wave	Wavelength range/m	Method of production	Properties and applications
Ultraviolet	10^{-10} to 10^{-8}	Electrons in atoms that were raised to high energy states by heat or by electric fields falling to lower permitted energy levels	Used in fluorescent lamps and for detecting forged banknotes. Stimulates the production of vitamin D in the skin to cause tanning; makes some materials fluoresce.
Visible light	4×10^{-7} to 7×10^{-7}		Light focused on the retina of the eye creates a visual image in the brain. Can be detected by chemical changes to photographic film; and electrical charges on the CCDs in digital cameras. Essential energy source for plants undergoing photosynthesis.
Infrared	10^{-7} to 10^{-3}		Radiated by warm bodies. Used for heating and cooking; and in thermal imaging devices.
Microwave	10^{-4} to 10^{-1}	High-frequency oscillators such as a magnetron; background radiation in space	Energy is transferred to water molecules in food by resonance at microwave frequencies. Used for mobile phone; and satellite communications.
Radio	10^{-3} to 10^{5}	Tuned oscillators linked to an aerial	Wide range of frequencies allows many signals to be transmitted. Groups of very large radio-telescopes can detect extremely faint sources in space.

Table 5

Exam tip

You may be required to identify the nature of an electromagnetic wave having calculated its wavelength or frequency. You should be aware of the wavelength (and frequency) ranges for each section of the spectrum.

Worked example

A transmitter sends out electromagnetic waves of frequency 9.3 GHz. Calculate the wavelength of these waves, and state the region of the electromagnetic spectrum to which they belong.

Answer

Using $v = f\lambda$:

$$\lambda = \frac{v}{f} = \frac{3.0 \times 10^8 \text{ m s}^{-1}}{9.3 \times 10^9 \text{ Hz}} = 3.2 \times 10^{-2} \text{ m}$$

These waves are part of the microwave section of the spectrum.

Intensity of light

The intensity of light is the brightness or strength of the radiation that is received by an observer. The **intensity** of a star is not a measure of how much radiation is emitted by the star per unit time, but of how much radiation is received on Earth per unit time.

Intensity is often referred to as 'radiation flux':

Knowledge check 8

State the regions of the electromagnetic spectrum in which the frequency of the radiation is:

a 10 GHz

b 3×10^{16} Hz

c 98 MHz

Knowledge check 9

State the regions of the electromagnetic spectrum in which the wavelength of the radiation is:

a 500 nm

b 900 nm

c 3.0×10^{-12} m

The **intensity** (radiant flux) is the power of the radiation falling perpendicularly onto one square metre of a surface.

$$\text{intensity} = \frac{\text{power received}}{\text{area}}$$

$$I = \frac{P}{A} \quad \text{unit: } W\,m^{-2}$$

Intensity depends not only on the power of the source, but also on the amount by which the beam spreads out between the source and the receiver. For example, the Sun emits radiation at a power of $4 \times 10^{26}\,W$, but this radiation is spread out in all directions and the intensity that actually reaches the Earth is just over $1\,kW\,m^{-2}$. By comparison, a $5\,mW$ laser emitting a beam of diameter $1.5\,mm$ can form an image of intensity around $3\,kW\,m^{-2}$.

A formula for how intensity depends on distance from the source can be derived. Consider a point source that is emitting radiation at power P uniformly in all directions. The points at a fixed distance, r, from the source form a spherical shell (i.e. surface of a sphere) of radius r and surface area $A = 4\pi r^2$. Therefore at these points the intensity will be $P/4\,\pi r^2$. Note that this relationship between I and d is an inverse square law.

Exam tip

The inverse square law means that if the distance of the source from the receiver is doubled then the intensity will decrease by four times.

Worked example

A panel of photovoltaic cells has an area of $3.2\,m^2$. When the Sun's rays fall normally on the panel (at $90°$ to the surface) the solar radiation flux is $400\,W\,m^{-2}$.

a Calculate how much energy falls on the panel every second:

   i   when the Sun's rays are normal to the surface

   ii when the Sun's rays fall at an angle of $30°$ to the surface

b If the panel has an emf of $24\,V$ and a maximum output current of $15\,A$, calculate the efficiency of the panel.

Answer

a i   power = radiation flux × area = $400\,W\,m^{-2} \times 3.2\,m^2 = 1300\,J\,s^{-1}$ (to 2 s.f.)

   ii The component of the solar flux that falls normally on the panel is
       $400\,W\,m^{-2} \times \sin 30° = 200\,W\,m^{-2}$
       Therefore, power = $200\,W\,m^{-2} \times 3.2\,m^2 = 640\,J\,s^{-1}$

b Electrical power output = $V \times I = 24\,V \times 15\,A = 360\,W$

   $$\text{efficiency} = \frac{\text{useful power output}}{\text{total power input}} \times 100\% = \frac{360\ W}{1300\ W} \times 100\% = 28\%$$

Exam tip

A common misunderstanding is the idea that the intensity of radiation falling on the Earth in winter is less than in the summer because the Earth in one hemisphere is further away from the Sun. The change in distance due to the inclination of the Earth is negligible compared with the distance from the Sun. It is the *angle* at which the radiation falls onto the surface that reduces the intensity.

Developments in photocell technology have led to solar panels with efficiencies approaching 40%, making the use of solar power as a renewable energy source more viable. It has been proposed that an array of reflecting panels in the Sahara desert could provide a high percentage of Europe's energy needs in the future.

Superposition and interference

Earlier in this section, displacement–distance graphs and a displacement–time graph for a particle oscillating within a wave were used to illustrate some basic wave definitions. Such graphs can also be used to explain the ideas of **phase** and **phase difference**, which will help in understanding the effects of superposition and interference that occur when similar waves meet at a point.

Knowledge check 10

The intensity of light from a lamp falling on a surface $0.50\,m$ away is $8.0\,W\,m^{-2}$. What will be the intensity of the light falling on the surface if the lamp is moved to $2.00\,m$ from the surface?

Consider the displacement–time graph in Figure 20. Suppose that the timing of the oscillation begins when the particle is moving upward through the midpoint — this is the point marked 0 on the graph. The points labelled 1, 2, 3 and 4 represent the stages, or **phases**, of one complete cycle of the vibration.

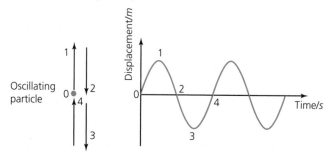

Figure 20

- Point 1 — the particle is at the positive extreme position. It has completed one quarter of a cycle and is said to be 90° or $\pi/2$ radians out of phase with point 0.
- Point 2 — the particle is moving down through the midpoint. It is exactly half a cycle behind point 0, which corresponds to a phase difference of 180° or π radians. Points 0 and 2 are said to be in antiphase.
- Point 3 — the particle is at the negative extreme position, three-quarters of a cycle from the starting point. This corresponds to a phase difference of 270° or $3\pi/2$ radians.
- Point 4 — the particle has reached the end of one cycle and is at the same position and moving in the same direction as point 0. The phase difference between point 4 and point 0 is 360° or 2π radians, and we say that the points are now back in phase.

From the displacement–distance graph in Figure 21, it can be seen that the phase of particles in the wave varies continuously along one complete wavelength. Hence the motion of particle A is in phase with that of particle E, but is π radians out of phase (i.e. in antiphase) with the motion of particle C. Similarly, B is in phase with F but in antiphase with D.

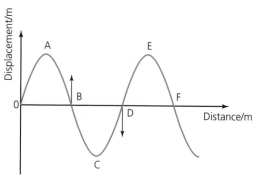

Figure 21

Worked example

A wave motion is described by the graphs in Figure 22.

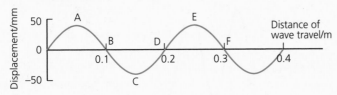

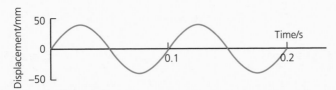

Figure 22

a State the amplitude.

b State the wavelength.

c Give all the pairs of points, other than (O and D) and (A and E), which are in phase.

d Give all the pairs of points, other than (O and B) and (A and C), which are in antiphase.

e Give all the pairs of points, other than (O and A) and (A and B), that have a phase difference of $\frac{\pi}{2}$.

f State the period of the motion.

g Determine the frequency of the motion.

h Hence calculate the wave speed.

Answer

a Amplitude = maximum displacement = 40 mm

b Wavelength λ = distance between O and D = 0.20 m

c B and F are the only other points that are in phase.

d (B and D), (C and E), and (D and F) are in antiphase.

e (B and C), (C and D), (D and E), and (E and F) differ in phase by $\frac{\pi}{2}$

f Period T = time for one oscillation = 0.10 s

g Frequency $f = \dfrac{1}{T} = \dfrac{1}{0.10 \text{ s}} = 10 \text{ Hz}$

h Wave speed $v = f\lambda = 10 \text{ s}^{-1} \times 0.20 \text{ m} = 2.0 \text{ m s}^{-1}$

Superposition

When two waves of the same type (e.g. two water waves) arrive at the same point, their displacements will combine to result in a wave of different amplitude.

If the waves are in phase, they add together to give a wave with an amplitude equal to the sum of the amplitudes of the original waves (Figure 23). This is called **constructive superposition**.

If the two waves are in antiphase (π radians out of phase), they 'cancel each other out' giving a resultant displacement of zero. This is called **destructive superposition**.

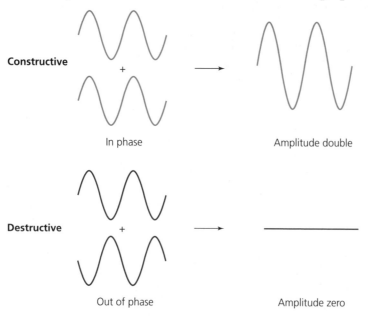

Figure 23

Wavefronts

You will probably have seen a ripple tank being used to demonstrate the properties of waves on a water surface (if not, there are plenty of simulations online that can be accessed by entering 'ripple tank' into a search engine). Waves can be observed moving across the surface as a series of lines (Figure 24).

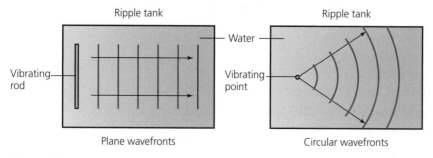

Figure 24

The darker lines represent the 'crests' of the wave and the light regions are at the 'troughs' where the water is shallower. The particles in the wave along a particular crest will all be in phase — such a line is known as a **wavefront**.

As a wave progresses, it can be seen that the wavefronts are always at right angles to the direction in which the wave is moving. Therefore, waves radiating out from a point source have wavefronts that are circular (or spherical in three dimensions), whereas plane wavefronts represent waves moving along parallel paths.

Exam tip

Superposition occurs only when two waves *of the same type* meet at a point.

A **wavefront** is a line joining points in a wave that are in phase with each other.

The progression of the wavefronts can be explained using **Huygens' construction**. Huygens considered every point on a wavefront as the source of secondary spherical wavelets that spread out with the wave velocity (Figure 25). The new wavefront is the envelope of these secondary wavelets.

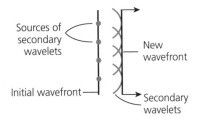

Figure 25 Huygens' construction

The distance between consecutive wavefronts is the wavelength of the wave in the medium through which it is travelling.

Coherence

Wave sources are said to be coherent if:

- the waves are of the same type
- the waves have the same frequency
- the sources are always in phase, or maintain a constant phase difference

Coherent sources can be obtained from the same wavefronts (i.e. secondary wavelets) or by connecting transmitters to the same oscillator (Figure 26).

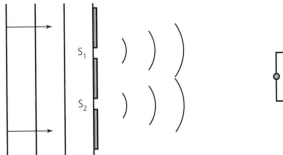
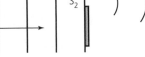
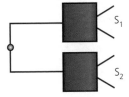

Figure 26

Conventional light sources, such as tungsten filament lamps, cannot be used as coherent sources because they emit photons of radiation (see the 'particle nature of light' section) with different wavelengths and random phase differences.

Lasers emit groups of photons that have the same frequency and are all in phase, so the emitted radiation is considered to be coherent.

Interference

When waves from two coherent sources meet at a point, they will have a constant phase relationship. For example, if the sources are constantly in phase with each other and the point is equidistant from both, then the waves will always arrive at the point exactly in phase (assuming that they passed through the same medium).

Constructive superposition will occur at the point, and a high-amplitude wave will be detected. Similarly waves meeting at a point that is one half-wavelength further away from one source than the other will always be out of phase at that point, so destructive superposition results.

Figure 27

If two loudspeakers are connected to the same output from a signal generator and placed about a metre apart in a large hall (Figure 27), a listener walking across the hall will detect a series of loud and quiet regions at regular intervals. This pattern of alternating maximum and minimum intensities is called an **interference pattern**.

Any pair of coherent sources can generate such interference patterns, including sources of water, radio and light waves (Figure 28).

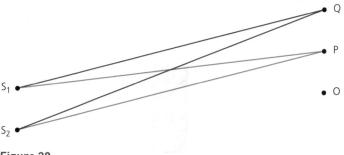

Figure 28

If the coherent sources S_1 and S_2 are in phase, the waves meeting at O will have travelled the same distance and so will be in phase and interfere constructively. So there will always be a wave of maximum intensity at this point.

At P, the wave from S_2 has travelled half a wavelength further than the distance travelled by the wave from S_1, so the waves interfere destructively at this point.

At Q, the distance from S_2 is a full wavelength longer than that from S_1 and, like at O, constructive interference takes place.

The difference between the distances from the two sources to a given point, $|S_2P - S_1P|$, is called the **path difference**.

For an interference pattern to be observed:
- two coherent sources are required
- the sources should be of similar amplitude

Exam tip

Note that superposition of waves does not necessarily lead to interference. Waves can meet randomly at a point and momentarily cause an increase or decrease in amplitude — but for a fixed pattern to occur the superposition needs to be continuous at that point.

If the sources are in phase, then:

- when the path difference is zero or equal to a whole number of wavelengths, constructive interference takes place
- when the path difference is an odd number of half-wavelengths, destructive interference occurs

Knowledge check 11

State the conditions required for an interference pattern to be formed.

Worked example

Two speakers are placed about 1 metre apart in a school hall. They are connected to the same signal generator so that the sounds from each are emitted in phase. A student standing at the opposite end of the hall, at a point equidistant from both speakers, hears a loud sound from the speakers. The student then walks across the hall parallel to the line from speaker to speaker and hears the sound intensity decrease and increase alternately. She makes a mark on the floor at the position of the third intensity minimum from her starting point. The distance of this point is measured to be 6.20 m from one speaker and 8.20 m from the other.

a Calculate the wavelength of the sound in air.

b If the frequency of the signal generator is 420 Hz determine the speed of sound in air.

c If the frequency is increased explain how the pattern of maxima and minima will be affected.

Answer

a Path difference = 8.20 m − 6.20 m = 2.00 m

For the third minimum the path difference is 2.5λ, so $\lambda = \dfrac{2.00 \text{ m}}{2.5} = 0.80$ m

b $v = f\lambda = 420 \text{ Hz} \times 0.80 \text{ m} = 336 \text{ m s}^{-1}$

c Increasing the frequency reduces the wavelength. The path difference that determines the positions of maxima and minima will therefore decrease, so the loud and quiet positions will become closer together.

Applications of interference

In addition to finding the frequency, wavelength and speed for a variety of waves, interference effects are used extensively in the optics industry. Lens surfaces can be ground to tolerances of around 500 nm with the aid of optical interferometers, which make use of small changes in the interference patterns produced by reflections of light from glass surfaces when the separation between the surfaces is altered.

Active noise control is now commonly used for protecting pilots and tractor drivers. The noise is recorded and electronically adjusted before being fed back into earpieces in such a way that it is out of phase with the original sound, leading to destructive interference. The noise is effectively cancelled out.

Knowledge check 12

Sketch a simple block diagram to illustrate the action of noise-control earphones.

Standing waves

Standing (stationary) waves are a particular example of interference. They are set up when two waves of equal frequency and amplitude that are travelling at the same speed in opposite directions superimpose.

Consider two coherent sources that are facing each other, an integral number of half-wavelengths apart (Figure 29). The same principles can be applied as for the production of an interference pattern described earlier.

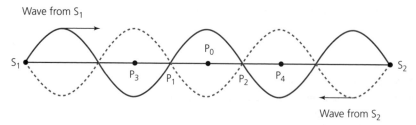

Figure 29

At the midpoint, P_0, between S_1 and S_2, the path difference from the sources is zero, so the waves will always be in phase and superimpose constructively. At this position, the air particles oscillate with maximum amplitude. However, at P_1 and P_2, the path difference is half a wavelength, so destructive interference takes place and the air particles at these positions always have zero amplitude. In other words, there is no movement of particles at P_1 or P_2 — such a point is called a **node**.

At points P_3 and P_4 the path difference equals one whole wavelength. These, like P_0, are therefore positions where the wave has maximum amplitude, called **antinodes** (Figure 30).

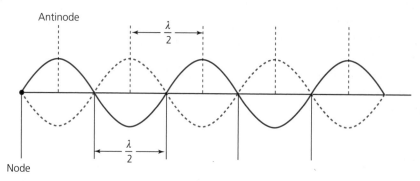

Figure 30

From Figure 30 you can see that the distance between adjacent nodes (and between adjacent antinodes) is half a wavelength.

A progressive wave moves by virtue of the phase difference between adjacent particles. In a standing wave, however, all the points between any two consecutive nodes are in phase at all times. Progressive waves transfer energy along the direction of wave travel, but in stationary waves energy transforms between kinetic energy and potential energy as the particles oscillate about their mean positions (except at the nodes, where the energy is zero).

Exam tip

Never refer to points of maximum and minimum amplitude on a progressive wave as antinodes and nodes. The terms only apply to stationary waves.

The differences between progressive and stationary waves are summarised in Table 6.

Progressive waves	Stationary waves
Energy is transferred in the direction of wave travel.	Energy is stored within each vibrating particle.
All points along the wave have the same amplitude.	Amplitude varies between a maximum value at the antinodes to zero at the nodes.
Adjacent points in the wave have a different phase relationship.	All points between each pair of consecutive nodes have a constant phase relationship.

Table 6

Worked example

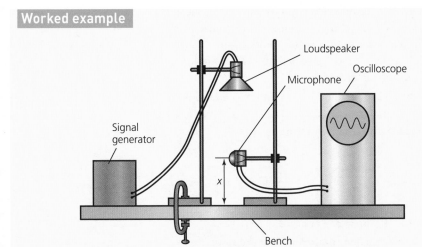

Figure 31

Figure 31 shows an experiment with sound waves. The height of the trace on the oscilloscope is proportional to the amplitude of the sound wave at the microphone. When the vertical distance, x, between the microphone and the bench is varied, the amplitude of the sound waves is found to change as shown in Figure 32.

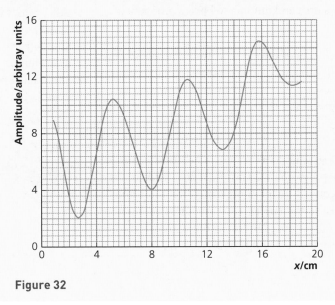

Figure 32

a Explain why the amplitude of the sound wave has a number of maxima and minima.

b The frequency of the sound waves is 3.20 kHz. Use this, together with information from the graph, to determine the speed of sound in air.

c The contrast between maxima and minima becomes less pronounced as the microphone is raised further from the surface of the bench. Suggest an explanation for this.

Answer

a The sound waves from the loudspeaker are reflected from the surface of the bench, giving rise to two identical waves travelling in opposite directions, which superimpose to form a stationary wave. Where the waves are in phase, an antinode of maximum intensity is formed; where the waves are in antiphase, a node of minimum intensity is formed.

b From the graph, the distance between the first and fourth minima (nodes) is
18.2 cm − 2.6 cm = 15.6 cm
Because the distance between adjacent nodes is $\frac{\lambda}{2}$ we have:

$$3 \times \frac{\lambda}{2} = 15.6 \text{ cm}$$

$$\lambda = \frac{15.6 \text{ cm} \times 2}{3} = 10.4 \text{ cm}$$

so $v = f\lambda = (3.20 \times 10^3 \text{ s}^{-1}) \times (10.4 \times 10^{-2})\text{m} = 333 \text{ m s}^{-1}$

c An ideal stationary wave, with all maxima and minima having the same amplitude, is formed from two travelling waves (in opposite directions) of the same amplitude. However, in this experimental situation the reflected wave loses some energy on reflection, and its amplitude decreases even further as it moves towards the loudspeaker. In contrast, the amplitude of the transmitted wave is largest near the loudspeaker. Consequently, the amplitudes of the transmitted and reflected waves become less similar as the microphone moves towards the speaker, causing the maxima and minima to be less pronounced.

As shown in the above example, measurements of wavelengths from the separations between nodes or antinodes can be used, with $v = f\lambda$, to determine frequency or wave speed. This example also demonstrates how interference of an incident wave and a reflected wave can create a standing wave.

Standing waves in strings

Figure 33 shows how standing waves in a wire under tension can be investigated. If the string is plucked at its midpoint, the disturbance travels to both ends of the string where it gets reflected. The reflections travel in opposite directions and superimpose to produce a standing wave in the string, with a node at each end and an antinode in the middle. Some of the energy in the string is transferred to the surrounding air, generating a sound of the same frequency as that of the string's vibrations.

Knowledge check 13

Calculate the wavelength of the sound from a trumpet if its frequency is 256 Hz and the speed of sound in air is 340 m s⁻¹

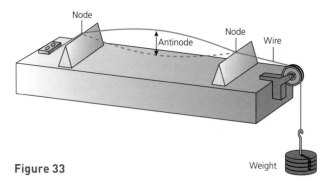

Figure 33

The speed of the wave in a string depends on the tension, T, that it is put under and the mass per unit length, μ, of the string and is given by the equation:

$$v = \sqrt{\frac{T}{\mu}}$$

The frequency of the sound produced by a vibrating string can be deduced by combining the above expression with the wave equation:

$$v = f\lambda = \sqrt{\frac{T}{\mu}} \quad \Rightarrow \quad f = \frac{1}{\lambda}\sqrt{\frac{T}{\mu}}$$

In the case of a string plucked at its midpoint, the length is the distance between two adjacent nodes of the standing wave and so equals half a wavelength, i.e. $\lambda = 2l$.

The frequency of a vibrating string is therefore given by the equation:

$$f = \frac{1}{2l}\sqrt{\frac{T}{\mu}}$$

Knowledge check 14

A wire of length 80 cm has a mass of 5.0 g. Calculate the speed at which a wave will travel along the wire if it is held taut by a force of 32 N.

Core practical 7

Investigate the frequency of a vibrating string or wire

Core practical 7 requires you to investigate the effects of changing the length, tension and mass per unit length on the frequency of a vibrating string or wire. Full details of how the measurements are made, the precautions to be taken and how the relationships can be represented in graphical form may be required for the examination.

The two most common experimental set-ups for investigating the factors affecting the fundamental frequency of a wire are shown in Figure 34.

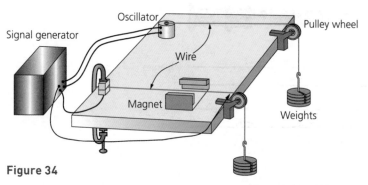

Figure 34

Several different standing waves can be produced in the same string. Because each end of the string is fixed these ends must be nodes. The standing wave with the longest possible wavelength is the one that has just a single antinode at the midpoint of the string — this wavelength corresponds to the **fundamental frequency** of the string. Other standing waves, with more nodes (N) and antinodes (A), can be set up as shown in Figure 35.

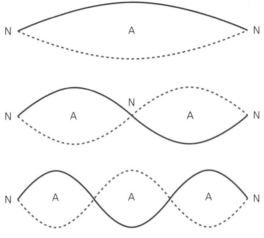

Figure 35

The standing waves with shorter wavelengths are called **overtones**, or **harmonics** if they have a frequency that is an integer multiple of the fundamental frequency.

The frequency of the sound emitted by a stringed instrument depends on the thickness and material used for the strings, along with their tension and length.

The tone of the instrument depends on the relative intensities of the harmonics.

Standing waves in air

The air column in a tube can be made to vibrate if a speaker or tuning fork is placed close to one open end of the tube.

Wind instruments also produce stationary waves. In a recorder, for example, there will be an antinode at both ends — the fundamental frequency can be varied (generating different musical notes) by opening or closing the stops with your fingers (Figure 36).

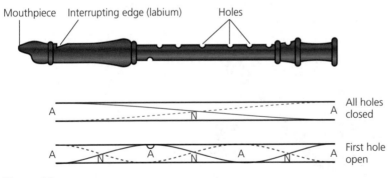

Figure 36

The air temperature can have an effect on the notes emitted from a wind instrument. As the temperature rises, the speed of sound in air increases. Using the wave equation $v = f\lambda$, you can see that for a standing wave of fixed wavelength, the frequency will increase, resulting in a higher-pitched note. This is one of the reasons why an orchestra tune their instruments on stage immediately before the performance.

Summary

After studying this section, you should be able to:
- define the terms amplitude, period, frequency, wave speed and wavelength of a wave, and use the wave equation $v = f\lambda$
- explain the difference between transverse and longitudinal waves
- recall the regions of the electromagnetic spectrum and be aware of the different properties, wavelengths and frequencies of each type of radiation
- understand the meaning of *radiation flux* (*intensity*)
- explain how two similar waves can be superimposed to produce a wave of larger or smaller amplitude
- define coherence and explain how two coherent sources can produce an interference pattern
- describe how stationary waves are formed, and determine the frequency and wavelength of such waves in strings and pipes

Refraction

Waves travel at different speeds in different media (materials). For example, the speed of sound is approximately $340\,\mathrm{m\,s^{-1}}$ in air at 20°C, $1500\,\mathrm{m\,s^{-1}}$ in water and $5000\,\mathrm{m\,s^{-1}}$ in steel; light travels at $3.0 \times 10^8\,\mathrm{m\,s^{-1}}$ in a vacuum, approximately $2.0 \times 10^8\,\mathrm{m\,s^{-1}}$ in glass and $2.3 \times 10^8\,\mathrm{m\,s^{-1}}$ in water.

When a wave is incident at an interface between two transmitting media, the change in speed can result in a change in the direction of travel of the wave. This effect is known as **refraction**.

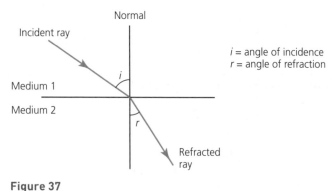

i = angle of incidence
r = angle of refraction

Figure 37

If the wave slows down as it passes from one medium into the next, it will deviate towards the normal, as shown in Figure 37. If the wave speeds up as it enters the second medium, it will deviate away from the normal.

The ratio of a wave's speed in the first medium to its speed in the second medium, denoted by $_1n_2$, is called the **refractive index** for that transmission.

A refractive index can be expressed in terms of the angles of incidence and refraction. This relationship is known as **Snell's law**.

> ### Exam tip
> The property of a material that relates to the wave speed through it is often referred to as the 'optical density'. The more optically dense a medium is, the slower the wave speed. So a wave travelling into a medium of higher optical density will slow down and bend towards the normal.

Worked example

A beam of ultrasound is passed through a fine membrane that separates air and carbon dioxide. The beam is incident in the air at an angle of 55° to the normal drawn at the interface and enters the carbon dioxide at an angle of 46° to the normal.

a Calculate the refractive index for sound travelling from air into carbon dioxide.

b Determine the speed of sound in carbon dioxide, given that the speed in air is $340\,\text{m s}^{-1}$.

Answer

a By Snell's law, $n = \dfrac{\sin 55°}{\sin 46°} = 1.14$

b Let v be the speed of sound in carbon dioxide. Then $1.14 = \dfrac{340\text{ m s}^{-1}}{v}$ and so

$$v = \frac{340\text{ m s}^{-1}}{1.14} = 300\text{ m s}^{-1}\ \text{(to 2 s.f.)}$$

Refraction of light

In many situations, light is refracted when it passes from air into glass, water or some other transparent medium. It is often useful to specify the **absolute refractive index** of a material for such refractions.

A useful way of expressing Snell's law in terms of absolute refractive indices is:

$$n_1 \sin \theta_1 = n_2 \sin \theta_2$$

where θ_1 is the angle of incidence in the first medium and θ_2 is the angle of refraction in the second medium.

> ### Exam tip
> The speed of light cannot exceed $3.0 \times 10^8\,\text{m s}^{-1}$ (the value in a vacuum). It follows that the refractive index for light passing through any medium must be greater than 1. If you calculate a value that is less than 1, it is likely that you have used the wrong angles.

The **refractive index** for a wave moving from one medium to another is the ratio of the wave speed in the first medium to that in the second:

$$_1n_2 = \frac{\text{speed in medium 1}}{\text{speed in medium 2}} = \frac{v_1}{v_2}$$

Snell's law states that the refractive index for a wave travelling from one medium to another is given by the ratio of the angle of incidence to the angle of refraction:

$$_1n_2 = \frac{\sin i}{\sin r}$$

where i is the angle of incidence and r is the angle of refraction.

The **absolute refractive index** (n) of a material is the ratio of the speed of light in a vacuum to its speed in the material.

$$n = \frac{\text{speed of light in a vacuum}}{\text{speed of light in the medium}}$$

Worked example

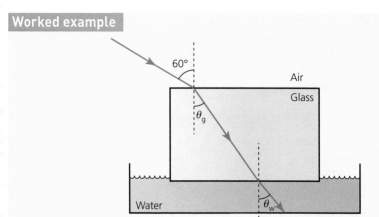

Figure 38

A ray of light enters a glass block at an angle of incidence of 60° and emerges through a layer of water on the opposite side (Figure 38). If the refractive index of glass is 1.55 and that of water is 1.33, calculate:

a the angle of refraction in the glass, θ_g

b the angle of refraction in the water, θ_w

Answer

a For the ray entering the glass block from above, we have $1.55 = \dfrac{\sin 60°}{\sin \theta_g}$.

   This gives $\sin \theta_g = \dfrac{\sin 60°}{1.55}$ and hence $\theta_g = 34°$

b For the ray passing from the glass block into the water underneath, we have:

   $$n_g \sin \theta_g = n_w \sin \theta_w$$

   $1.55 \sin 34° = 1.33 \sin \theta_w'$, giving $\theta_w = 41°$

Total internal reflection

When light travels from an optically dense (low speed) medium to a less dense (higher speed) medium (such as from glass into air) it deviates away from the normal (Figure 39).

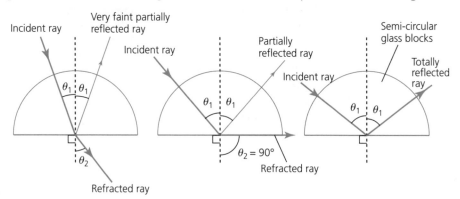

Figure 39

For small angles of incidence θ_1, most of the light will be refracted out of the glass block. As θ_1 is increased, the refracted ray will deviate further from the normal until it is eventually at 90° to the normal (i.e. θ_2 becomes 90°), parallel to the interface. For even larger angles of incidence, no light will be able to leave the glass and **total internal reflection** takes place. The angle at which this takes place is called the **critical angle**.

Applying Snell's law in this situation, we get:

$$n_1 \sin C = n_2 \sin 90°$$

$$\sin C = \frac{n_2}{n_1}$$

If the second medium is air, then n_2 is approximately 1, and $\sin C = \frac{1}{n_1}$.

> The **critical angle**, C, is the angle of incidence at which total internal reflection just takes place — it is the angle that leads to the light being refracted by exactly 90°.

Worked example

The refractive index of glass is 1.50 and that of water is 1.33. Calculate the critical angle for light passing from:

a glass to air

b water to air

c glass to water

Answer

a $\sin C = \frac{1}{n_g} = \frac{1}{1.50}$, so $C = 42°$

b $\sin C = \frac{1}{n_w} = \frac{1}{1.33}$, so $C = 49°$

c $\sin C = \frac{n_w}{n_g} = \frac{1.33}{1.50}$, so $C = 62°$

Applications of total internal reflection

In the food industry, the sugar concentrations of liquids can be measured from the refractive index of the solutions.

Reflections from silvered glass mirrors occur at both the top face of the glass and the bottom, silvered, face, producing a blurred image that is unsuitable for use in precision optical instruments. For reflections in binoculars, reflex cameras and the laser beam splitters of CD players, the inner surfaces of 45° prisms are used (Figure 40).

The most important recent developments in communications have been in the area of fibre optics. Glass fibres as thin as a human hair have a core surrounded by a cladding of a lower refractive index. This ensures that light passing through the core will be totally internally reflected as long as it impinges on the core–cladding boundary at an angle greater than the critical angle (Figure 40).

> **Exam tip**
>
> Remember that total internal reflection only occurs from dense to less dense boundaries (high refractive index to low refractive index).

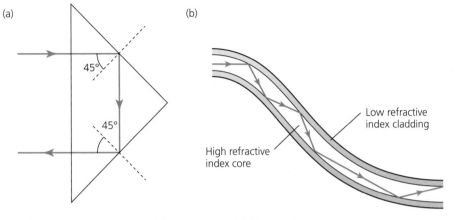

(a)   (b)

Knowledge check 16

Calculate the critical angle for light transmitted along an optical fibre, in which the refractive index of the core is 1.56 and that of the cladding is 1.48.

Figure 40 Light reflection in (a) a prism and (b) fibre optics

Worked example

A ray of light is directed at the midpoint of a semi-circular glass block. The angle of incidence is adjusted until total internal reflection just takes place. The critical angle is measured to be 41.5°.

The block is now put on a glass slide, with a layer of sugar solution between the block and the slide. The critical angle is now found to be 67.0°.

Calculate:

a the refractive index of the glass

b the refractive index of the sugar solution

Answer

a $\sin 41.5° = \dfrac{1}{n_g}$, so $n_g = \dfrac{1}{\sin 41.5°} = 1.51$ b $\sin 67.0° = \dfrac{n_s}{n_g} = \dfrac{n_s}{1.51}$, so $n_s = 1.51 \sin 67.0° = 1.39$

Lenses

Lenses focus light by refraction. The surfaces of lenses are shaped so that parallel rays of light passing through the lens will either converge to a single point or diverge from a single point. Figure 41 shows the passage of parallel rays through a **converging** and **diverging** lens.

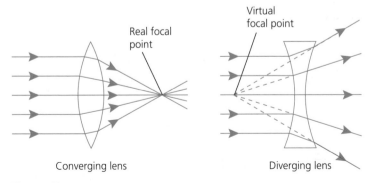

Real focal point

Virtual focal point

Converging lens   Diverging lens

Figure 41

The points where parallel rays meet or appear to diverge from are called the **focal points** of the lenses.

Figure 42 shows how a converging lens forms a **real** image where rays actually meet. The focal point for the diverging lens is on the same side as the source of the rays. A **virtual** image is the position from which the rays *appear* to originate, but no rays actually intersect at this point.

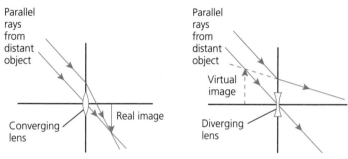

Figure 42

The **focal length** of a lens is the distance between the midpoint of the lens on the principal axis (the optical centre of the lens) and the focal point.

Ray diagrams

We can see objects because they give off light, either by direct emission or by scattering or reflecting light that falls on them. When some of this light passes through a lens an **image** of the object is formed. The image can be **real** or **virtual**.

A **real image** is formed by the actual intersection of rays of light and so can be projected onto a screen. A **virtual image** can only be seen when looking through the lens and appears to be at the point where the rays originate.

The positions, size and nature of the images produced by lenses can be deduced by drawing ray diagrams. There are three predictable rays from a point on an object that pass through a thin lens. These are shown in Figure 43, and can be used to determine the position of the image of that point.

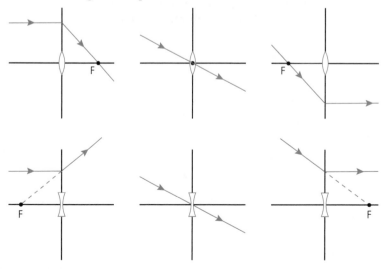

> The **focal point** of a lens is the point where parallel rays of light meet, or appear to diverge from, after passing through the lens.

Figure 43 Predictable behaviour of some rays of light

Ray diagrams are drawn using some or all of the predictable rays from a point on the object (usually the top) so that the position of the image of that point will be where the rays meet, or appear to originate from, after passing through the lens. It is common to represent the object as an arrow with its base on the **principal axis** and its head vertically above it. A ray of light from the foot of the arrow along the principal axis will pass through the **optical centre** of the lens without deviation, so it follows that the image of this point will be somewhere on the principal axis.

Power of lenses

The **power** of a lens relates to the ability of the lens to deviate rays of light through large angles. A powerful lens will have a focal point that is close to the centre — it will have a short **focal length**.

The power of a lens is not anything to do with to a rate of conversion of energy and is given the unit **dioptre** (m^{-1}). A lens of focal length $0.25\,m$ has a power of 4.0 dioptres.

If a combination of two or more lenses is used, the total power is the sum of the powers of the lenses:

$$P = P_1 + P_2 + P_3 +$$

The lens equation

The distance of an object from the optical centre of the lens (the object distance) is represented by u, the image distance by v and the focal length by f as shown in Figure 44.

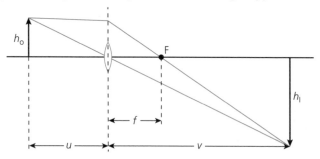

Figure 44

Using basic geometry, the following equations can be deduced:

$$\text{linear magnification} = \frac{\text{height of image}}{\text{height of object}} = \frac{h_1}{h_O} = \frac{v}{u}$$

$$\frac{1}{f} = \frac{1}{u} + \frac{1}{v}$$

This is called the lens equation and it can be used to work out the image position for any type of lens if the **real is positive** sign convention is used. All distances associated with real images and focal points are given positive values, and those associated with virtual images and focal points are negative.

> The **power** of a lens is the reciprocal of its focal length (in metres)
>
> $$P = \frac{1}{f}$$

Knowledge check 17

Calculate the combined power of two converging lenses; one having a focal length of 20 cm and the other of 25 cm.

Exam tip

The lens equation only applies to thin lenses. For ray diagrams and calculations the position of the lens is taken as a line perpendicular to the principal axis, passing through the optical centre of the lens.

Worked example

a Determine the position, nature and size of the image produced from an object of height 5.0 cm, placed: (i) 30 cm from a converging lens of focal length 20 cm; (ii) 40 cm from a diverging lens of focal length −50 cm.

b By considering the power of each lens, calculate the focal length of the two lenses combined together to form a single lens.

Answers

a i $\dfrac{1}{v} = \dfrac{1}{f} - \dfrac{1}{u} = \dfrac{1}{20 \text{ cm}} - \dfrac{1}{30 \text{ cm}} = \dfrac{1}{60 \text{ cm}}$

$v = +60 \text{ cm}$, so it is a real image

$m = \dfrac{v}{u} = \dfrac{60 \text{ cm}}{30 \text{ cm}} = 2.0$, so the height of the image is $2.0 \times 5.0 \text{ cm} = 10 \text{ cm}$

ii $\dfrac{1}{v} = \dfrac{1}{f} - \dfrac{1}{u} = \dfrac{1}{-50 \text{ cm}} - \dfrac{1}{40 \text{ cm}} = \dfrac{-9}{200 \text{ cm}}$

$v = -22 \text{ cm}$, so it is a virtual image

$m = \dfrac{v}{u} = \dfrac{-22 \text{ cm}}{-50 \text{ cm}} = 0.44$, so the height is $0.44 \times 5.0 \text{ cm} = 2.2 \text{ cm}$

b $P_1 = \dfrac{1}{+0.20 \text{ m}} = +5.0$ dioptres; $P_2 = \dfrac{1}{-0.50 \text{ m}} = -2.0$ dioptres

$P = P_1 + P_2 = (+5.0 \text{ dioptres}) + (-2.0 \text{ dioptres}) = +3.0 \text{ dioptres}$

So the combination is equivalent to a single converging lens of focal length 0.33 m (33 cm).

> **Knowledge check 18**
>
> Explain the difference between a real image and a virtual image.

Plane polarisation

In normal light waves, the electric vector oscillates in all directions within a plane perpendicular to the direction of wave travel. If such light is passed through a sheet of Polaroid®, the oscillations in all directions but one (i.e. the direction of the transmission axis) will be absorbed. The light emerging from the sheet has its electric vector oscillating in one direction only and is said to be **plane polarised**.

Suppose that another sheet of Polaroid® (or analyser) is held beyond the first sheet (the polariser). As this second analyser is rotated, alternating maximum and minimum (virtually zero) light intensities will be observed every 90°, depending on whether the transmission axis of the analyser becomes parallel or perpendicular to the transmission axis of the polariser. These phenomena are illustrated in Figure 45.

> **Plane polarised** waves are such that the particles, or fields, always oscillate in the same plane.

> **Exam tip**
>
> Longitudinal waves cannot be plane polarised because the oscillations are always parallel to the direction of wave motion.

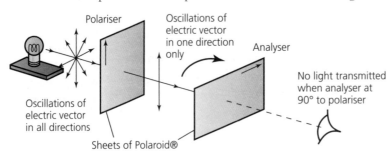

Figure 45

Microwaves can also be used to demonstrate polarisation. Microwave generators of the type found in school laboratories produce plane polarised waves of wavelength about 3 cm. The plane polarisation can be checked by using a suitable detector, such as an aerial, connected to an amplifier and an ammeter. When the aerial is parallel to the plane of polarisation, the maximum signal will be received. As the aerial is rotated about an axis joining it and the microwave transmitter, the signal intensity will diminish, reaching a minimum when the aerial has rotated through 90°.

Worked example

A microwave generator produces plane polarised waves of wavelength 30 mm.

a What is meant by a *plane polarised* electromagnetic wave?

b Draw a labelled diagram of the apparatus you would use to demonstrate that the microwaves are plane polarised.

c What is the frequency of these microwaves?

d What does this experiment show you about the nature of electromagnetic waves?

Answer

a A plane polarised electromagnetic wave is one in which the oscillations of the electric vector are in one direction only, in a plane perpendicular to the direction of wave travel.

b

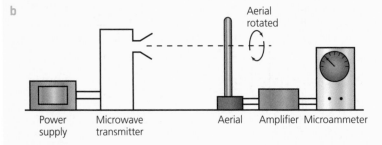

Figure 46

Exam tip

To gain full marks in an examination the diagram must be fully labelled and show precisely how the signal is detected, i.e. by an aerial, an amplifier and an ammeter (or by an aerial, an interface and a computer).

c Using $c = f\lambda$, where $c = 3.0 \times 10^8 \, \text{m s}^{-1}$ (because microwaves are electromagnetic waves), we have:

$$f = \frac{c}{\lambda} = \frac{3.0 \times 10^8 \, \text{m s}^{-1}}{30 \times 10^{-3} \, \text{m}} = 1.0 \times 10^{10} \, \text{Hz}$$

d Because the microwaves are plane polarised, this experiment shows that electromagnetic waves are transverse.

Exam tip

A grid of parallel copper wires can be used as an analyser. It is rotated and the aerial is kept fixed.

Knowledge check 19

Why is it not possible to polarise sound waves?

Applications of plane polarised light

Some complex molecules can rotate the plane of polarisation of transmitted light. In sugar solutions, the angle of rotation depends on the concentration of the solution.

With distilled water in the container shown in Figure 47, the analyser is rotated until its transmission axis is at right angles to that of the polariser, blocking out the light. A sugar solution is then poured into the container. Because the solution rotates the plane of polarisation, light will again be able to emerge. The degree of rotation of the plane of polarisation is measured by rotating the analyser until the light disappears again.

Perspex models of mechanical components are tested using stress analysis. When the model is loaded and viewed using light passed through a polariser and an analyser, a multicoloured stress pattern is revealed — this is due to rotation of the plane of polarisation by the strained regions. Potential areas of weakness can be detected from these observations.

Liquid crystal displays (LCDs) work by using electric fields to align crystals that rotate the plane of polarisation of light.

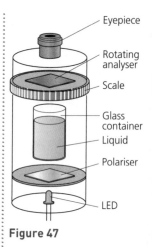

Figure 47

Diffraction

Diffraction occurs when a wavefront is disturbed by an obstacle or passes through an aperture. This is best illustrated by the water waves on a ripple tank (Figure 48).

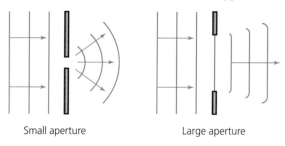

Small aperture   Large aperture

Figure 48

When the wavelength of a wave has the same order of magnitude as the width of the gap, the wavefront spreads out in an almost (semi)circular shape after passing through the gap, as the left-hand side of Figure 48 shows . This can be explained using Huygens' theory, with the secondary wavelets at each end of the wavefront 'spilling over' into the geometric shadow.

This explains why, even when someone is out of sight behind an open door, you can still hear the person talking — the wavelength of sound is about the same as the width of the door. Light, however, has a wavelength of the order of 500 nm, much narrower than the door. So it will undergo no discernable diffraction through the doorway.

For light, diffraction effects can be observed when the aperture is very small. If a laser beam is directed at the gap between the jaws of vernier calipers that are almost closed, a pattern can be seen on a screen beyond the gap. In addition to a central band

Diffraction is the spreading of waves after they have passed through an aperture, or have been disturbed by an obstacle.

Knowledge check 20

Explain why a radio signal may be received in some places in hilly regions, but a mobile phone is unable to pick up a signal from a transmitting dish that is close to the radio transmitter.

of maximum intensity there will be a series of bright and dark lines that arise from the effects of interference between the disturbed wavefronts (Figure 49). If the gap is narrowed, the central maximum will spread outwards.

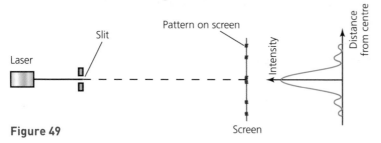

Figure 49

Diffraction patterns can also be observed when bright light is shone through fine materials such as net curtains.

Diffraction grating

When light is reflected from a surface with thousands of equally spaced, parallel grooves scored onto each centimetre — or transmitted through thousands of equally spaced, microscopic gaps — a diffraction pattern is produced. Such arrangements are called **diffraction gratings.**

Like a single slit, the width of the spacing must be of the same order of magnitude as the wavelength of light, but the patterns are different. The maxima occur at specific angles where the small, coherent waves from each groove or slit superimpose constructively producing sharply defined lines, as shown in Figure 50.

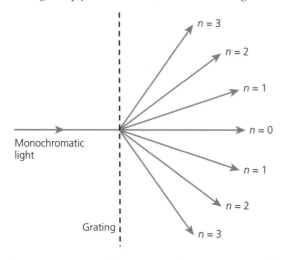

Figure 50 Diffraction grating

If the number of lines per metre on the grating is known it is possible to determine the wavelength of light transmitted or reflected by the grating by measuring the angles between the central maximum and the diffracted maxima. The relationship between the angles and the wavelength can be shown to be:

$$n\lambda = d \sin \theta$$

where d is the slit separation and n is the order of the maximum. It follows that for smaller values of d the value of θ will be bigger for a particular wavelength, and that the maximum number of orders of diffraction will be achieved when $n \leq d/\lambda$

Worked example

Green light from a laser pen is directed through a 500 lines mm^{-1} diffraction grating. The pattern on a screen placed 1.80 m behind the grating shows that the second order maximum is 1.13 m from the central maximum.

a Calculate the wavelength of the laser light.

b Determine the maximum order of the spectral maxima.

Answers

a $\tan\theta = \dfrac{1.13\ \text{m}}{1.80\ \text{m}} \rightarrow \theta = 32.1°;\ d = \dfrac{1}{500 \times 10^3\ \text{m}^{-1}} = 2.00 \times 10^{-6}\ \text{m}$

$2\lambda = 2.00 \times 10^{-6}\ \text{m} \times \sin 32.1° \quad \rightarrow \quad \lambda = 5.31 \times 10^{-7}\ \text{m} = 531\ \text{nm}$

b Maximum number of orders occurs when $n \leq \dfrac{d}{\lambda} = \dfrac{2.00 \times 10^{-6}\ \text{m}}{5.31 \times 10^{-7}\ \text{m}} \leq 3.77$

Because there can only be a whole number of orders, the maximum will be 3.

Core practical 8

Determine the wavelength of light using a diffraction grating

Core practical 8 requires you to determine the wavelength of light from a laser or other source using a diffraction grating. Full details of how the measurements are made, the precautions to be taken and how the wavelength is calculated may be required for the examination.

X-ray diffraction patterns, created by passing X-rays through crystal structures, were used to examine the atomic structures of many elements during the compilation of the periodic table. The spaces between the crystal layers behave like a three-dimensional diffraction grating. A similar method can be used to demonstrate the wave/particle nature of the electron (Figure 51).

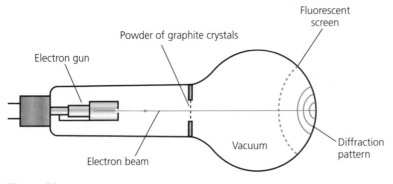

Figure 51

> **Exam tip**
>
> It must be stressed that diffraction effects are only observable when the obstacle or aperture is of the same order of magnitude as the wavelength.

Atomic separations in a graphite crystal (Figure 51) are of the order of 10^{-11} m, and electron diffraction suggests that the wavelength must be of the same order of magnitude. If the voltage across the tube is increased, transferring more energy to the electrons, then the rings of the diffraction pattern contract inwards indicating a reduction in wavelength. The relationship between the wavelength and the energy of electromagnetic radiation will be explored in more detail in the section on the particle nature of light.

Reflection

The laws of reflection of light state that:

- the angle of incidence equals the angle of reflection (Figure 52).
- the incident ray, the reflected ray and the normal at the point of incidence all lie in the same plane.

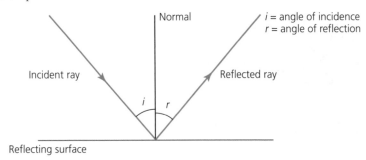

Figure 52

In this section more emphasis is put on normal reflections (reflections at right angles to the surface) occurring at interfaces between different media.

For light that is incident at right angles to a glass surface, some of it will be reflected from the surface while the rest will be transmitted (refracted) through the glass. The fraction of light that gets reflected depends on the relative refractive index going from air to glass.

For sound, the proportion that gets reflected from the interface between two media depends on a property known as *acoustic impedance*, which is related to the speed of sound in a medium and the density of the medium.

Pulse–echo techniques

The pulse–echo technique is a method for measuring the speed of a wave, or the distance from a reflecting surface, by measuring the time taken for a short pulse of radiation to reflect back to the original position. The method is commonly used to determine the speed of sound in air (Figure 53).

Figure 53

The operator starts the timing mechanism when the gun is fired, and then stops it when the echo is heard, recording the time t.

$$\text{speed} = \frac{\text{distance}}{\text{time}} = \frac{2x}{t}$$

If we know the speed v then the distance x can be calculated:

$$x = \frac{vt}{2}$$

Content Guidance

It is important to stress that, for this technique to work, the duration of the 'pulse' (the bang of the gun in this case) needs to be much shorter than the time taken for the wave to return. It would not be possible to time an echo from a source that emits a continuous sound.

In medicine, scans using ultrasound with frequencies on the order of 1 MHz can determine the depth of structures within the body (Figure 54).

Exam tip

Forgetting to include the return distance of the echo is a common error.

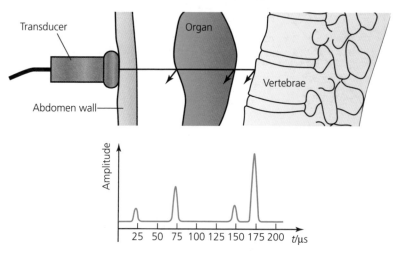

Figure 54

A pulse of ultrasound enters the body from the transducer. At each interface between different tissues in the body, a fraction of the sound gets reflected. These echoes are displayed on an oscilloscope that shows the time taken for each reflection to return to the transducer. The amplitude of each detected echo-signal represents the relative amount of ultrasound reflected at the corresponding interface.

Worked example

Use the timescale of the oscilloscope graph in Figure 54 to determine the distance of the organ from the inner abdomen wall and the thickness of the organ. The speed of sound is $1500\,\mathrm{m\,s^{-1}}$ in soft tissue and $1560\,\mathrm{m\,s^{-1}}$ in the organ.

Answer

Using $x = \dfrac{vt}{2}$, from the wall to the organ we have:

$$x = \frac{1500\,\mathrm{m\,s^{-1}} \times (75-25) \times 10^{-6}\,\mathrm{s}}{2} = 3.8 \times 10^{-2}\,\mathrm{m}$$

From the front to the back of the organ:

$$x = \frac{1560\,\mathrm{m\,s^{-1}} \times (150-75) \times 10^{-6}\,\mathrm{s}}{2} = 5.9 \times 10^{-2}\,\mathrm{m}$$

Knowledge check 21

Why must pulses of ultrasound be used in medical imaging?

Summary

After studying this section, you should be able to:

- describe how waves are refracted in terms of a change of wave speed across a boundary
- use Snell's law to determine relative and absolute refractive indices, and to calculate the critical angle at an interface
- know the meaning of *optical centre*, *principal axis*, *focal point* and *focal length* as applied to converging and diverging lenses
- understand the differences between *real* and *virtual* images
- determine the position, nature and size of images produced by converging and diverging lenses using ray diagrams and the lens equations
- be able to calculate the power of a lens from its focal length, and determine the power of lens combinations
- understand the concept of plane polarised waves, and describe how polarised light is produced and detected
- state the meaning of diffraction, and describe how diffraction is affected by the wavelength of the radiation and the size of the aperture or obstruction
- Use the diffraction grating formula $n\lambda = d\sin\theta$ to calculate the wavelength of light
- be aware that diffraction experiments show that electrons have wave properties
- use pulse–echo techniques to calculate distances or wave speeds

Particle nature of light

This section concentrates mainly on the photon nature of light and the development of basic quantum concepts. You will not be examined on the historical evolution of modern-day theories, but, if you have the time, a little background reading can be fascinating and may contribute to a fuller understanding of the nature of light. A brief synopsis of some of the significant stages is outlined below.

In the seventeenth century, Sir Isaac Newton and Christiaan Huygens proposed conflicting theories about the nature of light. Newton's corpuscular model was based on a study of the mechanics of particles, while Huygens suggested that light behaves as a wave. Around 1800, Thomas Young performed his famous double-slit experiment, which demonstrated the formation of interference patterns — a clear indication of the wave nature of light. When measurements of the speed of light in water were made later in the nineteenth century, the value turned out to be slower than the speed of light in air, providing further evidence that contradicted Newton's theory.

Although James Clerk Maxwell's equations, published in the late nineteenth century, linked the variations of electric fields and magnetic fields with the speed of light, there were still some phenomena that could not be explained using the classical wave model.

At the start of the twentieth century, Max Planck suggested that the reduction in peak wavelength of emissions of electromagnetic radiation at higher temperatures (e.g. the colour of hot objects glowing from red to orange to white as the temperature is increased) was consistent with the light being emitted in 'bundles' or 'quanta' of energy, with each package having an energy that depends on the frequency of the wave.

After Albert Einstein used Planck's ideas to explain the photoelectric effect and Niels Bohr adapted Rutherford's atomic model to predict spectral emissions, the photon nature of light became universally accepted.

Exam tip

It is important to understand that light, and many other electromagnetic radiations, consist of photons, and that the properties of the radiation depend on the energy of each photon.

The final link was made by Louis de Broglie, who showed that photons exhibit both wave properties and particle properties, indicating that Newton's theory was valid after all.

Planck's equation

A photon is a 'quantum' of wave energy. You can think of it as a short burst of electromagnetic waves emitted in a time of about 1 ns. A source of light gives out vast numbers of photons every second.

The energy carried by each photon is given by:

$$E = hf$$

where f is the frequency of the radiation and h is **Planck's constant**, with a value of 6.63×10^{-34} J s.

Worked example

Calculate:

a the energy in one photon of blue light of frequency 6.92×10^{14} Hz

b the wavelength of a photon with energy 3.44×10^{-19} J

Answer

a $E = hf = (6.63 \times 10^{-34} \text{ J s}) \times (6.92 \times 10^{14} \text{ s}^{-1}) = 4.59 \times 10^{-19}$ J

b $E = hf = h\dfrac{c}{\lambda}$

So $\lambda = \dfrac{hc}{E} = \dfrac{(6.63 \times 10^{-34} \text{ J s}) \times (3.00 \times 10^{8} \text{ m s}^{-1})}{3.44 \times 10^{-19} \text{ J}} = 5.78 \times 10^{-7}$ m

From this calculation, we find that the energy in a photon of blue light is about 4.6×10^{-19} J. This is a very small amount and it is often more convenient to use a non-SI unit called the **electronvolt (eV)**.

> An **electronvolt** is the work done on an electron when moving it through a potential difference of 1 volt.

Recall the definition of potential difference, from which we have $W = VQ$. So:

$$1\,\text{eV} = 1\,\text{V} \times (1.6 \times 10^{-19}\,\text{C}) = 1.6 \times 10^{-19}\,\text{J}$$

In many instances, photons are emitted when charges are accelerated by a potential difference, in which case the electronvolt is a particularly convenient energy unit. For example, in X-ray tubes electrons are accelerated towards the anode by voltages of around 100 kV (for diagnosis) or about 10 MV (for therapy). In such cases the energy of the electrons, and of the resultant X-rays, is typically given in keV or MeV.

Knowledge check 22

Calculate the photon energy of:

a radiation of frequency 7.2×10^{15} Hz

b radiation of wavelength 6.3×10^{-11} m

Worked example

Calculate the wavelength of a 120 keV X-ray.

Answer

$$E = 120 \times 10^{3}\,\text{eV} \times (1.6 \times 10^{-19}\,\text{J eV}^{-1}) = 1.92 \times 10^{-14}\,\text{J}$$

$$\lambda = \frac{hc}{E} = \frac{(6.63 \times 10^{-34}\,\text{J s}) \times (3.00 \times 10^{8}\,\text{m s}^{-1})}{1.92 \times 10^{-14}\,\text{J}} = 1.0 \times 10^{-11}\,\text{m}$$

Exam tip

When using Planck's equation $E = hf$ for calculations, the energy must always be converted to joules (by multiplying by 1.6×10^{-19}).

Photoelectric effect

The photoelectric effect refers to the emission of electrons from a material (usually a metal or a metallic oxide) when light is shone onto its surface. Classical wave theory was unable to explain why certain frequencies of light were able to release electrons, even at very low intensity, while other, lower, frequencies failed to eject any electrons regardless of the intensity of the incident radiation.

Einstein used Planck's equation and the law of conservation of energy to provide a simple explanation. The electrons require a minimum amount of energy, known as the **work function**, ϕ, to release them from the metal.

If a photon of incident light transfers its energy to an electron, the electron will be dislodged only if the photon energy is equal to or greater than the work function. If the photon energy is higher than the work function, the excess energy will be transferred as kinetic energy of the emitted electron. This can be expressed as follows:

energy of photon = work done to eject electron + kinetic energy of emitted electron

$$hf = \phi + \frac{1}{2}mv^2_{max}$$

> The **work function** of a material is the minimum amount of energy a photon requires to release an electron from the surface of the material.

Exam tip

This equation represents the case where all the residual photon energy is transformed to kinetic energy of the electron. In practice, some electrons may require extra energy to reach the surface, so their kinetic energy will be less than the maximum value given in the equation.

Different elements have different work functions. Only 1.9 eV is needed to release an electron of caesium, whereas 4.3 eV is needed for zinc and about 5.0 eV for copper.

The minimum frequency of incident radiation needed to release a photoelectron is called the **threshold frequency**, denoted by f_0. In this case, all the photon energy is used to liberate the electron, with none left over to transfer as kinetic energy. In Einstein's equation the kinetic energy term will be zero, so the work function ϕ equals hf_0.

Exam tip

Most common metals have work functions greater than the energy of photons of visible light, and so need ultraviolet radiation for photoelectric emission.

Worked example

When visible light is shone onto a polished zinc surface, no photoelectric emission is observed. However, electrons are emitted from the zinc when ultraviolet light of wavelength 200 nm is used. Explain why this occurs, given that the work function of zinc is 4.3 eV.

Answer

$$\phi = hf_0 = 4.3\,\text{eV} = 4.3\,\text{eV} \times (1.6 \times 10^{-19}\,\text{J\,eV}^{-1}) = 6.88 \times 10^{-19}\,\text{J}$$

$$f_0 = \frac{\phi}{h} = \frac{6.88 \times 10^{-19}\,\text{J}}{6.63 \times 10^{-34}\,\text{J\,s}} = 1.0 \times 10^{15}\,\text{Hz}$$

The shortest waves in the visible spectrum are at the blue end, with a wavelength of about 400 nm. From $c = f\lambda$ we find that the corresponding frequency is given by:

→

$$f = \frac{3.0 \times 10^8 \text{ m s}^{-1}}{400 \times 10^{-9} \text{ m}} = 7.5 \times 10^{14} \text{ Hz}$$

This is less than the threshold frequency for zinc, so no photoelectric emission occurs when visible light is shone onto the zinc surface.

For ultraviolet light, $f = \frac{3.0 \times 10^8 \text{ m s}^{-1}}{200 \times 10^{-9} \text{ m}} = 1.5 \times 10^{15} \text{ Hz}$.

This frequency is higher than f_0 and so the ultraviolet photons have sufficient energy to remove the electrons.

Knowledge check 23

Calculate the work function of a metal if the maximum kinetic energy of the photoelectrons emitted by a photon of energy 9.3×10^{-19} J is 1.3×10^{-19} J. Give your answer in eV.

The kinetic energy of photoelectrons can be found by measuring their **stopping potential**.

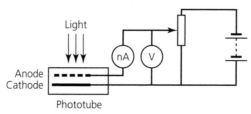

Figure 55

A phototube is an evacuated glass envelope containing an anode and a cathode (Figure 55). The cathode is coated with caesium so that when light is incident on it, photoelectrons will be emitted. A nanoammeter connected across the electrodes will detect a small current as the electrons move from the cathode to the anode. If a reverse potential difference is applied (the cathode is connected to the positive terminal of a battery) then the electrons will be slowed down. As the potential difference is increased, more work is done on the electrons; when the work done becomes equal to the maximum kinetic energy of the electrons they will stop moving, and the current on the ammeter will read zero. The minimum voltage required to stop all the electrons is called the stopping potential, denoted by V_s. Using the relationship $W = QV$ the minimum work done on an electron to stop it moving is eV_s, where e is the charge on a single electron, 1.6×10^{-19} C:

$$eV_s = \frac{1}{2} mv_{max}^2$$

Einstein's equation can now be written as:

$$hf = hf_0 + eV_s$$

Exam tip

If Einstein's equation is rearranged as $eV_s = hf - hf_0$, you can see that a graph of the maximum KE (eV_s) against the frequency of the incident light (f) will give a straight line of gradient h intersecting the energy axis at $-hf_0$. When $eV_s = 0$ you can see that $f = f_0$, and so the line will cross the frequency axis at the threshold frequency.

Worked example

Figure 56 shows how the maximum kinetic energy of photoelectrons emitted from the surface of sodium metal varies with the frequency, f, of the incident electromagnetic radiation.

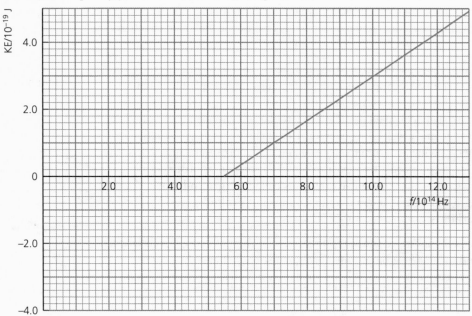

Figure 56

a Use the graph to find a value for the Planck constant.
b Use the graph to find the work function, ϕ, of sodium metal.
c Calculate the stopping potential when the frequency of the incident radiation is 7.0×10^{14} Hz.

Answer

a From $\frac{1}{2}mv^2_{\text{max}} = hf - \phi$ we can see that h is the gradient of the graph, so:

$$h = \frac{(4.9 - 0.0) \times 10^{-19}\,\text{J}}{(13.0 - 5.5) \times 10^{14}\,\text{s}^{-1}} =$$

$$= 6.5 \times 10^{-34}\,\text{J s}$$

b When $\frac{1}{2}mv^2_{\text{max}} = 0$, $hf_0 = \phi$, where $f_0 = $ threshold frequency $= 5.5 \times 10^{14}$ Hz

$$\phi = 6.5 \times 10^{-34}\,\text{J s} \times 5.5 \times 10^{14}\,\text{s}^{-1}$$

$$= 3.6 \times 10^{-19}\,\text{J}\ (= 2.3\,\text{eV})$$

c $\frac{1}{2}mv^2_{\text{max}} = hf - \phi$

$$= (6.5 \times 10^{-34}\,\text{J s} \times 7.0 \times 10^{14}\,\text{s}^{-1}) - 3.6 \times 10^{-19}\,\text{J}$$

$$= (4.6 - 3.6) \times 10^{-19}\,\text{J}$$

$$= 1.0 \times 10^{-19}\,\text{J} = eV_s$$

$$V_s = \frac{1.0 \times 10^{-19}\,\text{J}}{1.6 \times 10^{-19}\,\text{C}} = 0.63\,\text{V}$$

Exam tip

The KE can be checked directly from the graph — at a frequency of 7.0×10^{14} Hz the KE = 1.0×10^{-19} J.

Atomic spectra

When gases or vaporised elements are heated to high temperatures, or have a very large potential difference applied across them, they emit electromagnetic radiation consisting of a number of characteristic wavelengths. This group of frequencies is often shown as lines on a diffraction image, and is called the **emission spectrum** for that element.

Rutherford's model of an atom, consisting of a positive nucleus with electrons orbiting like the planets around the Sun, was a major step forward in explaining atomic behaviour. However, classical physics predicted that electrons ought to collapse inwards because such a charged particle in a circular orbit accelerates towards the centre of the orbit, and accelerating charges release electromagnetic radiation and lose energy. By applying quantum theory to a model of a hydrogen atom, Niels Bohr showed that an electron could only exist in certain discrete or **quantised** orbits. The lowest-energy state — the most stable orbit — is called the **ground state**. If energy is transferred to an electron, it can exist in a number of **excited** states, also called **permitted orbitals**.

The permitted orbitals are generally represented on a chart of energy levels labelled $n = 1, 2, 3$ and so on. The ground state corresponds to $n = 1$ and the energy differences between successive levels get smaller as n increases. Figure 57 shows the energy levels in a hydrogen atom.

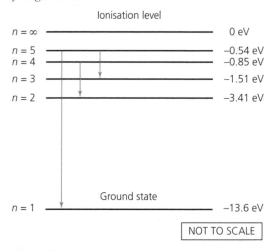

Figure 57

If sufficient energy is given to an electron, it can be completely removed from the atom (which becomes a positive ion). For a hydrogen atom, 13.6 eV is needed to raise the electron from its ground state to the ionisation level (labelled $n = \infty$). By convention, an electron that has just been removed from an atom is assigned 'zero' energy, so energy levels below ionisation take negative values — these indicate how much energy is needed to ionise the atom from each level.

When an electron falls from an energy level to a lower level, it releases the energy difference in the form of one quantum of radiation, hf. So when an electron falls from energy level E_2 to energy level E_1:

$E_2 - E_1 = hf$

An **emission spectrum** consists of all the characteristic frequencies emitted by the atoms of an element when they are at high temperatures or in a strong electric field.

The arrows in Figure 57 illustrate some possible transitions between energy levels. Electrons in higher energy levels can fall to any lower level. Different elements have characteristic differences between their atomic energy levels, therefore the emitted photons have a set of characteristic frequencies that can be used to identify the element via spectral analysis. For hydrogen, the energy emitted by an electron falling to level $n = 2$ will produce photons in the visible region, while more energetic, higher-frequency photons in the ultraviolet region will be released when an electron falls directly to the ground state.

Exam tip

Energy levels are often quoted in electronvolts and need to be converted to joules (multiplied by $1.6 \times 10^{-19}\,\mathrm{J\,eV^{-1}}$) in order to calculate the frequency of the emission using Planck's equation.

Worked example

Use the values in Figure 57 to answer the following questions about hydrogen atoms.

a Calculate the ionisation energy in joules for an electron in the $-13.6\,\mathrm{eV}$ level.

b What is the wavelength of the light emitted when an electron falls from the $-1.51\,\mathrm{eV}$ level to the $-3.41\,\mathrm{eV}$ level? Suggest what colour the light would be.

c Between which energy levels must an electron fall to emit blue light having a wavelength of 434 nm?

d Without doing any calculations, explain why the radiation emitted when an electron falls to its lowest energy level cannot be seen.

Answer

a Ionisation energy $= 13.6\,\mathrm{eV} = 13.6 \times (1.6 \times 10^{-19})\mathrm{J} = 2.2 \times 10^{-18}\,\mathrm{J}$

b $hf = E_2 - E_1 = (-1.51)\mathrm{eV} - (-3.41)\mathrm{eV} = 1.90\,\mathrm{eV} = 1.90\,\mathrm{eV} \times (1.6 \times 10^{-19}\,\mathrm{J\,eV^{-1}}) = 3.04 \times 10^{-19}\,\mathrm{J}$

$$f = \frac{3.04 \times 10^{-19}\,\mathrm{J}}{6.63 \times 10^{-34}\,\mathrm{Js}} = 4.59 \times 10^{14}\ \mathrm{Hz}$$

$$\lambda = \frac{c}{f} = \frac{3.00 \times 10^8\,\mathrm{m\,s^{-1}}}{4.59 \times 10^{14}\,\mathrm{s^{-1}}} = 6.54 \times 10^{-7}\ \mathrm{m} = 654\ \mathrm{nm}$$

This falls in the range of visible red light.

c $$f = \frac{c}{\lambda} = \frac{3.00 \times 10^8\ \mathrm{m\,s^{-1}}}{434 \times 10^{-9}\ \mathrm{m}} = 6.91 \times 10^{14}\ \mathrm{Hz}$$

$E_2 - E_1 = hf$

$= (6.63 \times 10^{-34}\,\mathrm{Js}) \times (6.91 \times 10^{14}\,\mathrm{s^{-1}}) = 4.58 \times 10^{-19}\,\mathrm{J}$

$$= \frac{4.58 \times 10^{-19}\ \mathrm{J}}{1.60 \times 10^{-19}\ \mathrm{J\,eV^{-1}}} = 2.86\ \mathrm{eV}$$

This can be achieved by the electron falling from the $-0.54\,\mathrm{eV}$ level to the $-3.41\,\mathrm{eV}$ level.

d The energy released when an electron falls to the lowest level would be much higher than the value of 2.86 eV calculated above, which corresponds to blue light. The radiation emitted must therefore have a much higher frequency and lie beyond the visible region of the electromagnetic spectrum.

Knowledge check 24

Explain the terms:

a permitted orbital

b quantum jump

c ionisation energy

Wave–particle duality

The wave nature of an electron was mentioned in the diffraction section earlier. Photons of light are deviated by huge gravitational forces as they pass close to black holes, and 'radiation pressure' can be measured for a stream of photons. These observations suggest that certain 'particles', such as electrons, have wave properties, while 'waves', such as photons of light, can behave like particles.

Louis de Broglie introduced a relationship between wave behaviour and particle behaviour by suggesting that particles with momentum p would have a corresponding wavelength λ given by:

$$\lambda = \frac{h}{p}, \text{ where } h = \text{Planck's constant} = 6.63 \times 10^{-34} \text{ J s}$$

It would appear that Newton's corpuscular theory of light had some merit after all.

Worked example

a Calculate the wavelength of a 10 keV electron. (Take the mass of an electron to be 9.1×10^{-31} kg.)

b In which part of the electromagnetic spectrum is this wavelength?

Answer

a $10 \text{ keV} = 10 \times 10^3 \text{ eV} \times 1.6 \times 10^{-19} \text{ J eV}^{-1} = 1.6 \times 10^{-15} \text{ J}$

$$\frac{1}{2} mv^2 = 1.6 \times 10^{-15} \text{ J}$$

$$v^2 = \frac{2 \times 1.6 \times 10^{-15} \text{ J}}{9.1 \times 10^{-31} \text{ kg}}$$

$$v = 5.9 \times 10^7 \text{ m s}^{-1}$$

$$\lambda = \frac{h}{p} = \frac{h}{mv} = \frac{6.63 \times 10^{-34} \text{ J s}}{9.1 \times 10^{-31} \text{ kg} \times 5.9 \times 10^7 \text{ m s}^{-1}} = 1.2 \times 10^{-11} \text{ m}$$

b This is in the X-ray region of the electromagnetic spectrum.

Summary

After studying this section, you should be able to:

- understand that light consists of photons, and that the energy of a photon is given by the expression $E = hf$
- recollect that photon energy is often expressed in *electron volts* (eV), and that 1 eV is equivalent to 1.6×10^{-19} J
- use Einstein's equation to explain the photoelectric effect and use it to determine values for the *work function, threshold frequency, maximum electron energy and stopping potential*

- explain how emission spectra are produced and calculate the frequencies of photons emitted when an electron falls from an excited state to a lower energy permitted-orbital
- be aware that photons have both particular properties and wave properties and that these are linked by de Broglie's equation

Questions & Answers

The Edexcel examinations

This guide covers only the sections on materials and waves and the particle nature of light that are required for AS paper 2 and for part of A-level papers 2 and 3.

The Edexcel AS level examination consists of two papers, each containing multiple choice, short open, open response, calculations and extended writing papers. Both papers are of 1 hour and 30 minutes duration and have 80 marks. The examination is intended for students who have completed a 1-year course of study that is based on the core physics of the A-level specification.

The A-level examination consists of three papers. Paper 1 and paper 2 containing multiple choice, short open, open response, calculations and extended writing questions. Both papers are of 1 hour and 45 minutes duration and have 90 marks.

Paper 3 covers the general and practical principles of physics and is of 2 hours and 30 minutes duration and has 120 marks. This paper covers all of the topics and includes synoptic questions as well as assessing the conceptual and theoretical understanding of experimental methods.

Note that both papers 1 and 2 at AS and A-level will also examine 'Working as a Physicist'. Briefly this means students:

- working scientifically, developing competence in manipulating quantities and their units, including making estimates
- experiencing a wide variety of practical work, developing practical and investigative skills by planning, carrying out and evaluating experiments and becoming knowledgeable of the ways in which scientific ideas are used
- developing the ability to communicate their knowledge and understanding of physics
- acquiring these skills through examples and applications from the entire course.

A formulae sheet is provided with each test. Copies may be downloaded from the Edexcel website, or can be found at the end of past papers.

Examiners use certain **command terms** that require you to respond in a particular way. You must be able to distinguish between these terms and understand exactly what each requires you to do. A full list can be found in the Edexcel specification Appendix 7.

You should pay particular attention to diagrams, sketching graphs and calculations. Many candidates lose marks by failing to label diagrams properly, by not giving essential numerical data on sketch graphs and, in calculations, by not showing all the working or by omitting the units in calculations.

About this section

The two tests that follow are made up of questions similar in style and content to the AS and A-level examinations. The first test is in the style of an AS paper 2, but section B is restricted to the topics in this guide.

You may like to attempt a complete paper in the allotted time and then check your answers, or maybe do the multiple-choice section and selected questions to fit your revision plan. It is worth noting that there are 80 marks available for the 90 minute test, so this should help in determining how long you should spend on a particular question. You should therefore be looking at about 10 minutes for the multiple-choice section and just over a minute a mark on the others.

The second test is in the style of the A-level paper 2, but again is restricted to the topics covered in this guide. There are 90 marks available and the time allotted is 1 hour and 45 minutes.

You should also be aware that during the examination you must write your answers directly onto the paper. This will not be possible for the tests in this book, but the style and content are the same as the examination scripts in every other respect. It may be that diagrams and graphs that would normally be added to on the paper, have to be copied and redrawn. If you are doing a timed practice test, you should add an extra few minutes to allow for this.

The answers should not be treated as model answers — they represent the bare minimum necessary to gain the marks. In some instances, the difference between an A-grade response and a C-grade response is suggested. This is not possible for the multiple-choice section, and many of the shorter questions do not require extended writing.

Ticks (✓) are included in the answers to indicate where a mark has been awarded. Half marks are not given.

■AS Test Paper

Time allowed: 1 hour 30 minutes. Answer ALL the questions.

Section A

For Questions 1–8 select one answer from A to D.

Question 1

The units of tensile strain are:

A $N\,m^{-1}$ **B** $N\,m^{-3}$ **C** no unit   **D** Pa   (1 mark)

Questions 2 and 3 relate to the following displacement–distance graph of a progressive wave.

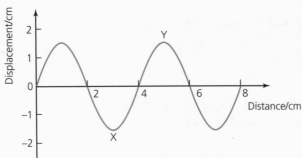

Question 2

The amplitude of the wave is:

A 1.5 cm   **B** 2.0 cm   **C** 3.0 cm   **D** 4.0 cm   (1 mark)

Question 3

The phase difference between points X and Y on the wave is:

A $\frac{\pi}{4}$ radians   **B** $\frac{\pi}{2}$ radians   **C** π radians   **D** 2π radians   (1 mark)

Question 4

The pressure at the bottom of a column of liquid of height 80 cm and density 900 kg m⁻³ is about:

The pressure at the bottom of a column of liquid of height 80 cm and density $900\,kg\,m^{-3}$ is about:

A 7.2 Pa   **B** 72 Pa   **C** 720 Pa   **D** 7.2 kPa   (1 mark)

Question 5

The diagram shows how an image is formed when an object is placed a small distance away from a thin converging lens.

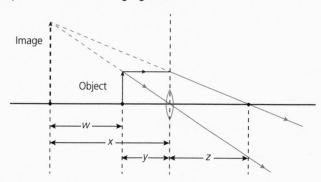

Which of the labels w, x, y or z represents the focal length of the lens?

A w **B** x **C** y **D** z (1 mark)

Question 6

An object is placed at a distance of 12 cm from a thin converging lens of focal length 10 cm. The image of the object, formed by the lens will be:

A real and erect

B real and inverted

C virtual and erect

D virtual and inverted (1 mark)

Questions 7 and 8 relate to the following diagram, which shows some of the energy levels in a mercury atom.

```
0  ——————————————————  Ionisation

−1.6 ————————————————


Energy/eV
  −5.5 ————————————————



−10.4 ———————————————  Ground state
```

Question 7

The ionisation energy for an electron in the ground state is:

A 2.6×10^{-19} J   **B** 7.8×10^{-19} J   **C** 8.8×10^{-19} J   **D** 1.7×10^{-18} J   (1 mark)

ⓔ To ionise an atom the electron must be displaced from the ground state to the ionisation level.

Question 8

The wavelength of the photon emitted when an electron falls from the –1.6 eV level to the –5.5 eV level is:

A 120 nm   **B** 230 nm   **C** 250 nm   **D** 320 nm   (1 mark)

Total: 8 marks

ⓔ Convert the change in energy levels to joules using $\Delta E = hf$ and $c = f\lambda$.

Answers to Questions 1–8

1 C

ⓔ Strain is the ratio of extension (m) and original length (m) and so has no units.

2 A

ⓔ The amplitude is measured from the midpoint to the peak = 1.5 cm

3 C

ⓔ X and Y are half a cycle out of phase; phase difference = π radians

4 D

ⓔ $\Delta P = h\rho g = 0.80$ m $\times$ 900 kg m^{-3} $\times$ 9.8 m s^{-2} = 7056 Pa $\approx$ 7.2 kPa

5 D

ⓔ A ray of light that is parallel to the principal axis will always pass through the focal point after passing through the lens.

6 B

A ray diagram shows that the image of the top of the object is formed by the intersection of rays on a screen will form a real, inverted image

7 D

ⓔ Energy needed for ionisation is 10.4 eV.

10.4 eV = 10.4 eV $\times$ (1.6 $\times$ 10^{-19} J eV^{-1}) = 1.7 $\times$ 10^{-18} J

8 D

ⓔ $\lambda = \dfrac{hc}{E} = \dfrac{6.63 \times 10^{-34} \text{ J s} \times 3.00 \times 10^{8} \text{m s}^{-1}}{[-1.6-(-5.5)] \text{ eV} \times 1.6 \times 10^{-19} \text{J eV}^{-1}} = 3.2 \times 10^{-7}$ m = 320 nm

Question 9

Waves on the sea enter a harbour and are diffracted as they pass through the harbour gate.

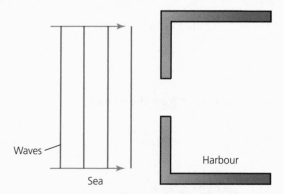

This question tests your knowledge of diffraction and how wavefronts and wavelength are affected by obstructions and wave speed.

(a) Explain the meaning of diffraction. (1 mark)

(b) Copy the diagram, and add three more wavefronts of the wave entering the harbour. (2 marks)

ⓔ Three wavefronts must be drawn, all separated by the same distance (same as the waves before they enter the harbour).
A typical grade-C response may show circular wavefronts spreading out but have them closer together, farther apart or with variable wavelength after entering the harbour.
Take care with these diagrams – sloppy efforts lose marks!

(c) Further up the coast, as the waves approach the shore, the wavelengths get shorter. What does this tell you about the speed of the waves in shallower water? (1 mark)

Total: 4 marks

ⓔ This is just a one mark answer. Use the equation $v = f\lambda$ to relate speed and wavelength.

Student answer

(a) Diffraction is the spreading out of waves when they pass through an aperture or are obstructed by an object. ✓

(b) The drawing should show three wavefronts with a (semi)circular shape having the same wavelength (equal spacing between consecutive wavefronts) as in the open sea ✓, and spreading out by at least 45° ✓.

(c) The waves travel more slowly in the shallower water. ✓

Question 10

The following is a passage from a report on the advantages and disadvantages of using a stainless steel kettle compared with using a plastic one. Complete the gaps in the passage by selecting appropriate words from this list:

brittle   dense   elastic   harder   ions

plastic   polymers   stronger   tougher

If the plastic kettle is dropped, or receives a sharp blow, it is more likely to crack or shatter than the steel one. This is because the material is (i) _____. The steel is much (ii) _____ , and is capable of absorbing much more energy, and may suffer from only minor indentations caused by (iii) _____ deformation. The surface of the steel is (iv) _____ than the plastic and so less likely to be scratched.

The plastic kettle will retain its heat much better. It consists of long-chain molecules called (v) _____ , which gives it a low thermal conductivity. The plastic is also less (vi) _____ than the steel and so the kettle is lighter.

Total: 3 marks

ⓔ This question tests your knowledge of the properties of solid materials. High marks can be scored if the details shown in Table 4 on p.18 are studied.

Answers		
(i) brittle	**(iii)** plastic	**(v)** polymers
(ii) tougher	**(iv)** harder	**(vi)** dense
All six correct: 3 marks; 4 or 5 correct: 2 marks; 2 or 3 correct: 1 mark.		

Question 11

A child dips a wire frame into a soap solution to blow some bubbles. She notices that there are multicoloured patterns on the film when it is held up to the light.

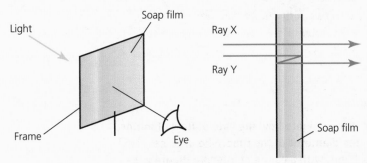

The right-hand part shows a cross-section of the soap film with two adjacent rays passing through a certain region. Ray X passes straight through the film, but ray Y undergoes two reflections on the inner surfaces before emerging. In this region the thickness of the film is about 450 nm. The wavelength of blue light in the soap solution is about 300 nm.

(e) You need to be aware that interference will take place between the two rays. An A-grade student will be able to relate the path difference and the wavelength when constructive interference occurs.

(a) Explain why this region appears blue to the child. (3 marks)

(b) The soap film increases in thickness from top to bottom. Suggest a reason for this. (1 mark)

(c) When a red lamp is viewed through the film, a series of bright and dark horizontal stripes is seen. Explain this effect, and determine the minimum thickness of the soap film for a dark line to appear if the wavelength of red light in the soap solution is 500 nm. (3 marks)

Total: 7 marks

Student answer

(a) The reflected ray travels backwards and forwards between the surfaces. The path difference between this ray and the one passing straight through is 2×450 nm = 900 nm. ✓

The path difference is equal to three complete wavelengths of blue light in the solution. ✓

So the two waves emerge in phase and constructive superposition occurs for blue light. ✓

(b) Gravitational force pulls the liquid down; or the solution flows down, making the film thicker at the bottom. ✓

(c) When the path difference is a whole number of wavelengths, constructive interference occurs; when the path length differs by an odd number of half-wavelengths, destructive interference takes place. ✓

The minimum thickness t for destructive interference is the thickness for which $2t = \lambda/2$ ✓; that is, $t = \lambda/4 = 125$ nm ✓

(e) Producing destructive interference by two reflections that give rise to a path difference of half a wavelength is seen commonly in examination questions. Examples include reflections from the edges of the 'bumps' on CDs and non-reflective coatings on lenses.

Question 12

A student uses a simple viscometer to investigate how the rate of flow of motor oil through a glass tube depends on the diameter of the tube. She uses several pieces of tube having the same length, but with a range of internal diameters.

A diagram of the apparatus used by the student and the graph of her results is given here.

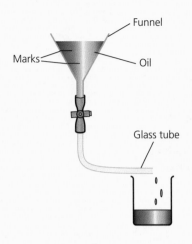

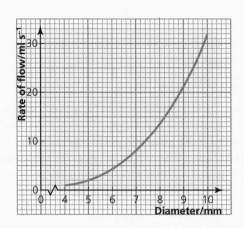

(a) Use the graph to work out the rate of flow of oil for tubes of diameter 5 mm and 10 mm.

(2 marks)

(b) Suggest a possible relationship between the rate of oil flow and the internal diameter of the tubes.

(2 marks)

(c) State two other factors that will affect the rate of flow of the oil through a glass tube.

(2 marks)

Total: 6 marks

ⓔ This question tests knowledge of the factors affecting the viscosity of a fluid. In part (b) an A-grade candidate will be aware that if doubling the diameter increases the flow 16 times, a fourth power relationship is a possibility. Candidates who simply state that the flow rate increases will not gain any marks, but a C-grade candidate can gain 1 of the marks by referring to the values from the graph.

Student answer

(a) When $d = 5.0$ mm the flow rate $= 2.0$ ml s^{-1} and when $d = 10$ mm the flow rate $= 32$ ml s^{-1}

One value $\pm\frac{1}{2}$ of a scale division $(2.0 \pm 0.5$ or $32 \pm 0.5)$ ✓

Both values correct including units. ✓

(b) The rate of flow through the larger diameter pipe is (about) 16 times that of the smaller one. ✓

This suggests that the flow rate is proportional to the fourth power of the diameter. ✓

(c) Other factors that may affect the flow rate are:
- pressure (height of funnel)
- temperature (affects the viscosity)
- length of tube

Any two ✓ ✓

Question 13

Some transparent materials are optically active and can rotate the plane of polarisation of light passing through them.

(a) Explain the meaning of plane of polarisation. (1 mark)

A polarimeter is a device for measuring the rotation of polarised light after it has passed through an optically active liquid. A simple version is shown below. It consists of a light source beneath a fixed polarising film (the polariser), together with a second, similar sheet of Polaroid that can be rotated around an angular scale (the analyser). A glass container holding a fixed volume of an optically active liquid is placed between the polariser and the analyser.

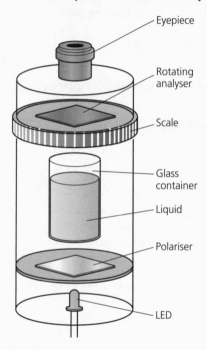

(b) Describe the procedure you would follow to determine the angle of rotation for a given sample of the liquid. (3 marks)

A low-calorie drink contains artificial sweeteners that are optically active. When a sample of the drink is placed in the polarimeter, the angle of rotation is found to be 40°.

The polarimeter is calibrated by measuring the angle of rotation for a range of concentrations of the sweetener. The results are given in the table.

Concentration/%	Angle of rotation/°
10	15
15	30
20	46
25	60

(c) Plot a graph of angle of rotation against concentration, and use this to determine the concentration of sweetener in the drink. (4 marks)

ⓔ Marks are awarded for correctly labelled axes, appropriate choice of scale, accurate plotting (± half a division) and drawing the line of best fit. Lines drawn on the graph (using a ruler) from the appropriate angle to the line, and from there to the concentration axis, will reduce the possibility of error in the final answer.

(d) Give one further application of optical activity. (1 mark)

Total: 9 marks

Student answer

(a) The plane of polarisation is the plane in which the vibrations of a (transverse) wave take place. ✓

(b) Without any solution in the container, adjust the analyser until the light is blocked out. ✓
Take the scale reading (it should be zero). ✓
With a sample in place, rotate the analyser again until the light vanishes. Note the scale reading and so determine the angle through which the plane of polarisation has been rotated. ✓

(c) The graph should have: the axes labelled 'concentration/%' and 'rotation/°' ✓; suitable scales (e.g. 0–30% and 0–80°) ✓; four points plotted and a straight line of best fit drawn ✓

The concentration of sweetener in the sample is estimated to be 18–19% ✓

ⓔ A similar marking scheme is followed for most graphs. It is best to copy the labels for the axes directly from the given table (e.g. 'concentration/%') so that the units are always included. The scale should be such that the given range of values fills most of the available space. Sometimes marks are awarded separately for the plotted points and the line or curve through them (especially when a curve is expected); but in this question, the points and the line both need to be drawn accurately to gain the 1 mark.

(d) Stress patterns in Perspex models, liquid crystal displays etc.

ⓔ A different example to the one in the question is needed. A candidate giving another example of liquid concentration would not gain this mark.

Question 14

A schematic diagram showing the use of an ultrasound A-scan to measure the diameter of a baby's head is shown below. The graph represents the reflections from the front and the rear of the skull as shown on a cathode ray oscilloscope monitor.

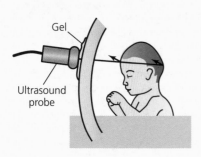

(a) What is ultrasound? (1 mark)

(b) A physics examination paper asked the question: 'Describe how an A-scan is used to measure the diameter of a fetal skull, and explain why a coupling gel is used between the probe and the skin.'

One student's answer was as follows:

Ultrasound is directed at the skull and reflected back to the probe. The reflections are shown as two peaks on the screen. The distance between the peaks represents the diameter of the skull. The coupling gel is needed to replace the air between the probe and the skin because the ultrasound will reflect back off the air.

Discuss the student's answer, highlighting any incorrect or missing physics. (5 marks)

ⓔ The 'student's answer' given in the question illustrates many of the frequently encountered errors and omissions made by AS and A-level candidates. Such an answer would gain zero marks from a possible 4.

(c) If the average speed of sound in soft tissue is 1560 m s⁻¹, use the information on the graph to determine the diameter of the baby's head. (3 marks)

Total: 9 marks

ⓔ Remember to include the return distance in the pulse–echo calculation.

Student answer

(a) Ultrasound is sound that has a frequency higher than that which can be heard by the human ear (usually higher than 20 kHz). ✓

(b) Pulses of ultrasound are needed. ✓

The cathode ray oscilloscope shows the time delay (not the distance) between the reflection from the front of the skull and the reflection from the rear. ✓

The distance is then calculated using speed × time. ✓

There is no mention that the time measured is the time needed for the pulse to travel to the rear of the skull and back again. So the depth of the skull is $\frac{vt}{2}$ ✓

(e) In addition to the above marks, one of the following can gain a fifth mark:

■ The ultrasound does not reflect off the air; it is almost totally reflected at the air–skin boundary. ✓
■ There is a big difference in the acoustic impedances of air and soft tissue. The acoustic impedance of the gel is similar to that of skin, allowing less of the ultrasound to be reflected and more to be transmitted. ✓

(e) The most common omission is the mention of 'pulses' – a continuous stream of ultrasound will mix up all the reflections and show nothing.

(c) Using $\text{speed} = \dfrac{\text{distance}}{\text{time}}$ or $\text{distance} = \text{speed} \times \text{time}$ ✓

gives: $d = \dfrac{vt}{2} = \dfrac{(1560\,\text{ms}^{-1}) \times (150 \times 10^{-6}\,\text{s})}{2}$ ✓ $= 0.12\,\text{m}$ ✓

Question 15

A photocell is a device that can be used to measure the intensity of light incident on it.

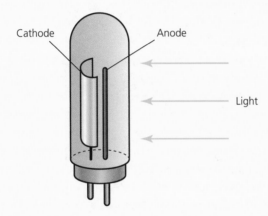

When photons of light hit the cathode, some electrons may be released. This emission depends on the energy of the photon and the work function of the cathode.

(e) This question requires the understanding and use of Einstein's photoelectric equation.

(a) Explain the meaning of 'work function', and state the conditions under which photoelectric emission will take place. (2 marks)

(b) Caesium has a work function of 1.9 eV. Calculate the lowest frequency of radiation that will produce photoelectric emission from a caesium-coated cathode. (3 marks)

(c) Visible light has a range of wavelengths from 400 nm to 700 nm. Explain, with appropriate calculations, whether or not a photocell with a caesium-coated cathode would be suitable for the whole of the visible range. (3 marks)

(e) 'Appropriate calculations' means that both extreme frequencies must be calculated and compared with the threshold frequency.

(d) A caesium cathode is exposed to the full range of wavelengths of the visible spectrum. Show that the maximum kinetic energy of the photoelectrons is about 1 eV.

(3 marks)

ⓔ Because this is a 'show that' question, the result (and the calculation leading up to it) must be given to two significant figures because the value stated in the question had only one significant figure.

(e) Give one reason why some photoelectrons will be emitted with kinetic energy lower than 1 eV.

(1 mark)

Total: 12 marks

Student answer

(a) The work function of a material is the minimum energy that is needed to remove an electron from atoms close to the surface. ✓

Photoelectric emission will happen if the energy of an incident photon is equal to or greater than the work function. ✓

(b) $1.9\,\text{eV} = 1.9 \times 1.6 \times 10^{-19}\,\text{J}$ ✓ $= 3.04 \times 10^{-19}\,\text{J}$

Using $\phi = hf_0$, $3.04 \times 10^{-19}\,\text{J} = 6.63 \times 10^{-34}\,\text{J s} \times f_0$ ✓

so $f_0 = 4.6 \times 10^{14}\,\text{Hz}$ ✓

(c) For red light: $f = \dfrac{c}{\lambda} = \dfrac{3.00 \times 10^8\,\text{m s}^{-1}}{700 \times 10^{-9}\,\text{m}} = 4.3 \times 10^{14}\,\text{Hz}$ ✓

For blue light: $f = \dfrac{3.00 \times 10^8\,\text{m s}^{-1}}{400 \times 10^{-9}\,\text{m}} = 7.5 \times 10^{14}\,\text{Hz}$ ✓

A photocell with a caesium-coated cathode would not cover the whole of the visible range, because visible light with frequencies from $4.3 \times 10^{14}\,\text{Hz}$ to $4.6 \times 10^{14}\,\text{Hz}$ will not be able to release a photoelectron. ✓

ⓔ The first mark will be given for using $f = \dfrac{c}{\lambda}$ with values correctly substituted for either the longest or the shortest wavelength, but correct values for both of the frequencies are needed to gain the second mark.

(d) Maximum photon energy $= hf_{max} = (6.63 \times 10^{-34}\,\text{J s}) \times (7.5 \times 10^{14}\,\text{Hz})$ ✓

$= 4.97 \times 10^{-19}\,\text{J} = \dfrac{4.97 \times 10^{-19}}{1.6 \times 10^{-19}}\,\text{eV} = 3.1\,\text{eV}$ ✓

$hf_{max} = \phi + KE_{max} \Rightarrow KE_{max} = 3.1\,\text{eV} - 1.9\,\text{eV} = 1.2\,\text{eV}$ ✓

(e) Some of the electrons further away from the surface may transfer some energy in the process of reaching the surface. The less energetic photons in the visible light range (those associated with the longer wavelengths towards the red end of the spectrum) will release electrons with lower kinetic energy. ✓

ⓔ Either of these reasons would gain the mark.

Section B

Question 16

Read the following press release and then answer the questions that follow it.

"A scientist, whose paper was deemed so hopeless that he had to pursue it in his spare time, has won the 2014 Nobel prize in physics for an invention that paved the way for widespread energy-efficient lighting.

Shiju Nakaruma shares the prize with Isamu Alasaki and Hiroshi Amano of Japan for 'the invention of efficient blue light emitting diodes, which has enabled bright and energy saving white light sources'.

Conventional light bulbs are inefficient because they work by heating up a wire filament. The hot filament produces light, but wastes substantial amounts of energy through lost heat.

In an LED, light is produced when negative electrons combine with positive 'holes' in wafer-thin layers of semiconductors. Red LEDs became widely available in the 1960s and adorned calculators and digital watches through the 1970s. Green LEDs were developed around the same time. But despite these early successes, scientists failed in their attempts to create blue LEDs.

These are crucial if white LED lighting was ever to become a reality because only blue light – which has the highest visible frequencies – can be converted into white light."

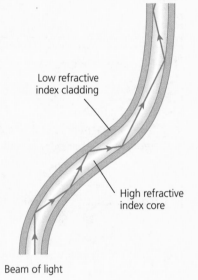

Low refractive index cladding

High refractive index core

Beam of light

e Section B will always include a passage from a scientific publication. You will be expected to use your knowledge and understanding of physics to answer the questions relating to the passage. In some cases the questions may apply to any of the core physics topics.

(a) What is meant by 'white light'? (1 mark)

(b) The energy of a photon emitted by a blue LED is 2.20 eV.
Calculate the frequency of the blue light. (2 marks)

(c) Explain why only the light from blue LEDs can be converted into white light. (3 marks)

e To answer part (c) you should consider the photon energies of red, green and blue light.

The manufacturer's specifications state that the power of a green LED laser pointer is 5 mW and the wavelength of the light is 530 nm. The pointer is used to transmit a beam of light through an optical fibre as shown in the diagram.

The refractive index of the core for green light is 1.62 and that of the cladding is 1.54.

(d) Calculate the critical angle between the core and the cladding for the light, and describe how the beam passes though the fibre. (3 marks)

(e) On leaving the fibre, the beam forms a circular image of diameter 4 mm on a plain wall. Calculate the intensity of the image on the wall. State any assumption you have make. (3 marks)

Total: 12 marks

Student answer

(a) White light is a mixture of all of the wavelengths in the visible region of the electromagnetic spectrum. ✓

ⓔ It is acceptable to state that white light can be produced by combining rays of red, green and blue light.

(b) $2.20\,\text{eV} = 2.20\,\text{eV} \times 1.6 \times 10^{-19}\,\text{J eV}^{-1} = 3.5 \times 10^{-19}\,\text{J}$ ✓

$$E = hf \Rightarrow f = \frac{E}{h} = \frac{3.52 \times 10^{-19}\,\text{J}}{6.63 \times 10^{-34}\,\text{J s}} = 5.3 \times 10^{14}\,\text{Hz} \checkmark$$

(c) Blue photons have a higher energy (hf) than green or red photons. ✓

In order to produce white light a full range of energies is needed. ✓

Blue photons have sufficient energy to release lower energy photons (e.g. from a phosphor) but it is not possible for a green photon or a red photon to release a higher energy photon. ✓

(d) $\sin C = \dfrac{n_{\text{cladding}}}{n_{\text{core}}} = \dfrac{1.54}{1.62} \Rightarrow C = 72°$ ✓

The ray is always incident on the core–cladding interface at an angle larger than 72° ✓ so total internal reflection takes place and no light escapes from the fibre. ✓

(e) $I = \dfrac{P}{A} = \dfrac{5 \times 10^{-3}\,\text{W}}{\pi \times (2 \times 10^{-3}\,\text{m})^2}$ ✓ $= 400\,\text{W m}^{-2}$ ✓

It is assumed that all of the energy is transferred through the fibre (none is absorbed or scattered by the glass); OR the value of the power given by the manufacturer is accurate. ✓

Question 17

A student carries out an investigation into the behaviour of some copper wire, by clamping one end of the wire to the bench and applying varying loads to the other.

She sets up the equipment as shown in the diagram, with a 3.00 m length of wire, and measures the extensions of the wire for a range of loads up to 40.0 N.

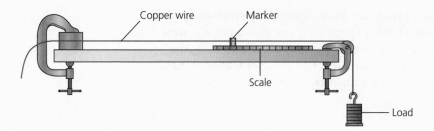

ⓔ This question relates to **Core practical 5** and the behaviour of a length of wire loaded up to its breaking point.

(a) Why is it beneficial to use such a long piece of wire? (1 mark)

(b) The student intends to analyse her results using a stress–strain graph.

    (i) What additional measurement does she need to make in order to determine the stress?

    (ii) Define the terms 'stress' and 'strain'. (3 marks)

ⓔ For part (b)(i) a common error is stating that the cross-sectional area is measured. This area is required but is calculated from the diameter. Similarly, 'radius' would not gain the mark.

(c) A typical graph for such an experiment is drawn below. Use the graph to determine the Young modulus and the ultimate tensile stress (UTS) of copper. (3 marks)

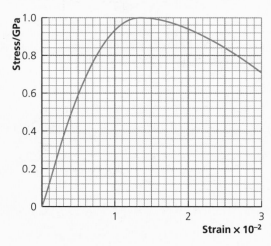

ⓔ In an examination, where the answers are written on the paper, an A-grade candidate would extend the linear region of the graph so that a more accurate value of the gradient is obtained from the larger triangle.

(d) As an extension to the investigation, the experiment is repeated using a steel wire of the same dimensions as the original wire, and then using an equal length of thinner copper wire. Stress–strain graphs are drawn using the same scales as in the first investigation. Describe the appearance of these graphs compared with that from the initial experiment.

(3 marks)

Total: 10 marks

Total for paper: 80 marks

Student answer

(a) A longer wire means a bigger/more readable extension ✓ (for the same load).

(b) (i) The diameter of the wire ✓

  (ii) $stress = \dfrac{force}{cross\text{-}sectional\ area}$ ✓; $strain = \dfrac{extension}{original\ length}$ ✓

(c) Young modulus = the gradient in the Hooke's law region

  $= \dfrac{0.6 \times 10^{9}\ Pa}{0.5 \times 10^{-2}} = 1.2$ ✓ $\times 10^{11}\ Pa$ ✓

  UTS = 1.0 GPa ✓

(d) A stress–strain curve for steel will have a steeper linear region ✓ (because its Young modulus is bigger than copper) and will reach a much bigger stress ✓ (before it begins to yield).

The curve for the thinner copper wire will be identical to the first graph ✓ (stress–strain graphs show the properties of the material regardless of the dimensions of the sample, because these are incorporated in calculating stress and strain).

■ A-level Test Paper

Time allowed: 1 hour 45 minutes. Answer ALL the questions.

Question 1

Which of the following waves cannot be polarised?

A light **B** microwaves **C** sound **D** X-rays (1 mark)

Question 2

Which of the following does not affect the rate of flow of a fluid through a pipe when a constant pressure is applied across the ends of the pipe?

A the density of the fluid

B the length of the pipe

C the radius of the pipe

D the temperature of the fluid (1 mark)

Question 3

Coherent sources must always have the same...

A amplitude **B** frequency **C** intensity **D** phase (1 mark)

In questions 4 and 5, choose the appropriate line on the stress–strain graph that best describes the property of the material.

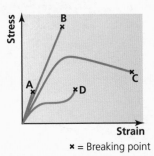

× = Breaking point

Question 4

A material that is strong and brittle (1 mark)

Question 5

A material that is tougher than the others (1 mark)

Question 6

The spike P on the screen of a CRO represents the pulse of sound from a SONAR transmitter fitted to the hull of a ship. The spike R is that of the same pulse that has been reflected from the ocean bed.

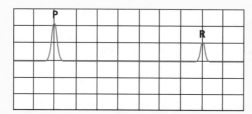

If the speed of sound in water is $1500\,m\,s^{-1}$ and the time base of the oscilloscope is set at $5\,ms$ per division, the depth of the water beneath the ship is:

A 6.5m   B 13m   C 26m   D 52m   (1 mark)

Question 7

The critical angle for light travelling from glass of refractive index 1.53 to water of refractive index 1.33 is:

A 32°   B 41°   C 49°   D 60°   (1 mark)

Question 8

Three identical springs, each having a spring constant of $60\,N\,m^{-1}$, are connected to support a load as shown in the diagram.

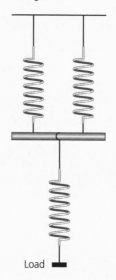

The spring constant of the combined springs is:

A $20\,N\,m^{-1}$ B $40\,N\,m^{-1}$ C $60\,N\,m^{-1}$ D $80\,N\,m^{-1}$ (1 mark)

When light of wavelength 599 nm is shone onto a caesium surface, photoelectrons are emitted. The work function of caesium is 1.9 eV. Use these data to answer questions 9 and 10.

Question 9

The threshold frequency of caesium is:

A 1.3×10^{-33} Hz   **B** 3.0×10^{-19} Hz   **C** 4.6×10^{14} Hz   **D** 2.9×10^{33} Hz   (1 mark)

Question 10

The maximum kinetic energy of the photoelectrons is:

ⓔ You need to calculate the photon energy (in eV) and then subtract the work function

A 0.18 eV   **B** 1.9 eV   **C** 2.1 eV   **D** 4.0 eV   (1 mark)

Total: 10 marks

Answers to Questions 1–10

1 C

ⓔ Sound waves are longitudinal waves, and so cannot be polarised.

2 A

ⓔ The density has no effect. The temperature affects the viscosity, and hence the flow rate.

3 B

ⓔ The frequency must be the same for coherent sources; usually coherent sources are also in phase, but this is not a requirement — the phases may be different provided that there is a constant relationship between them so D is not valid.

4 B

ⓔ This line has the highest ultimate breaking stress and so is the strongest; it has no plastic deformation so it is brittle.

5 C

ⓔ There is a long region of plastic deformation and the area under the curve (energy absorbed) is the greatest, and so it represents the toughest material.

6 C

ⓔ $\Delta t = 7.0\,\text{div} \times 5.0\,\text{ms div}^{-1} = 35\,\text{ms}$; $2x = 1500\,\text{m s}^{-1} \times 35 \times 10^{-3}\,\text{s}$; $x = 26\,\text{m}$

7 D

ⓔ $\sin C = \dfrac{n_w}{n_g} = \dfrac{1.33}{1.53} \Rightarrow C = 60°$

Questions & Answers

8 B

ⓔ For a given force, F, if the single spring extends by x, the double spring extends by $0.5x$. The total extension is therefore $1.5x$.

The spring constant of the combination is $\dfrac{F}{1.5x} = \dfrac{k}{1.5} = 40\,\mathrm{N\,m^{-1}}$

9 C

ⓔ $hf_0 = \phi = 1.9\,\mathrm{eV} = 1.9 \times 1.6 \times 10^{-19}\,\mathrm{J}$

So $f_0 = \dfrac{1.9 \times 1.6 \times 10^{-19}\,\mathrm{J}}{6.63 \times 10^{-34}\,\mathrm{J\,s}} = 4.6 \times 10^{-14}\,\mathrm{Hz}$

10 A

ⓔ $E = hf = \dfrac{hc}{\lambda}$

Energy of the incident photon $= \dfrac{(6.63 \times 10^{-34}\,\mathrm{J\,s}) \times (3 \times 10^{8}\,\mathrm{m\,s^{-1}})}{599 \times 10^{-9}\,\mathrm{m}}$

$= 3.32 \times 10^{-19}\,\mathrm{J} = \dfrac{3.32 \times 10^{-19}\,\mathrm{J}}{1.6 \times 10^{-19}\,\mathrm{J\,eV^{-1}}} = 2.08\,\mathrm{eV}$

So the maximum KE $= 2.08\,\mathrm{eV} - 1.90\,\mathrm{eV} = 0.18\,\mathrm{eV}$

Question 11

A standing wave is set up in a string as shown in the diagram. The frequency of the supply is 60 Hz.

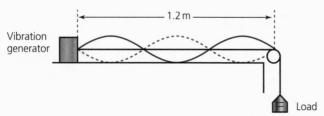

(a) Sketch the vibrating string, and label the position of one node and one antinode. (1 mark)

(b) What is the wavelength of the standing wave? (1 mark)

(c) The frequency of the vibration generator is altered until a standing wave with one more node is set up in the string. Calculate the frequency of the vibration generator when this occurs. (2 marks)

Total: 4 marks

ⓔ You can gain full marks for this question if you know the standard diagrams for standing waves.

> **Student answer**
>
> **(a)** A node should be labelled at one of the positions of minimum disturbance, and an antinode labelled at the midpoint of one of the loops. ✓
>
> **(b)** There are one and a half wavelengths on the string: $1.5\lambda = 1.2\,\mathrm{m}$, so $\lambda = 0.80\,\mathrm{m}$ ✓

ℯ A common mistake that candidates make with standing waves is to take the wavelength as the distance between two adjacent nodes or antinodes, instead of twice this length (i.e. the separation between every other node or every other antinode).

(c) With the addition of one node, there will now be two complete wavelengths along the string, so the wavelength will be 0.60 m ✓

Using $v = f\lambda$, as the speed of the wave on the string is unchanged, we have $f_1\lambda_1 = f_2\lambda_2$ and hence $f_2 = \dfrac{0.80 \text{ m} \times 60 \text{ Hz}}{0.60 \text{ m}} = 80 \text{ Hz}$ ✓

Question 12

The table below shows some features of the electromagnetic spectrum. Complete the table by filling in the missing entries (a), (b), (c) and (d).

Radiation	Typical wavelength	Source
Visible light	(a)	Very hot objects
X-rays	(b)	(c)
(d)	3.0 cm	Very high frequency oscillator (klystron tube)

Total: 4 marks

ℯ This question tests your knowledge of the properties of electromagnetic radiation. Full marks can be earned if the details in Table 5 are known.

Student answer

(a) [value between] 400–700 nm ✓

(b) [value between] 10^{-14} m to 10^{-10} m ✓

(c) (electrons striking) tungsten anode ✓

(d) microwaves ✓

Question 13

The diagram shows some of the energy levels of a hydrogen atom.

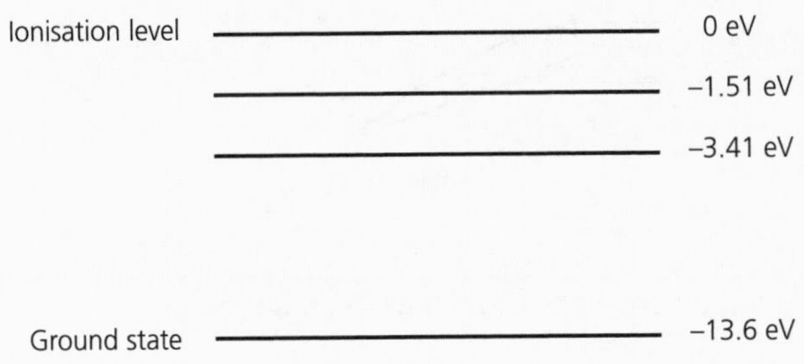

Ionisation level ———————————— 0 eV

———————————— −1.51 eV

———————————— −3.41 eV

Ground state ———————————— −13.6 eV

(e) This question is about the absorption and emission of energy when electrons are raised and lowered between permitted 'energy levels'.

(a) Calculate the ionisation energy in joules for an electron in the −13.6 eV level. (2 marks)

(b) A neutron of kinetic energy 12.9 eV collides with a hydrogen atom. As a result, an electron in the ground state is raised to the −1.51 eV level. What is the kinetic energy of the neutron after the collision? (1 mark)

(e) This mark will be lost if only the energy needed to raise the electron to the higher orbital is calculated. The question requires the determination of the residual energy of the neutron.

(c) A transition between which two energy levels will give rise to an emission of wavelength 654 nm? (3 marks)

Total: 6 marks

Student answer

(a) Ionisation energy = $13.6\,\text{eV} \times 1.6 \times 10^{-19}\,\text{J eV}^{-1}$ ✓ $= 2.2 \times 10^{-18}\,\text{J}$ ✓

(b) Energy gained by electron = $-1.51\,\text{eV} - (-13.6\,\text{eV})$

$= 12.1\,\text{eV}$ = energy lost by neutron

Kinetic energy of neutron after collision = $12.9\,\text{eV} - 12.1\,\text{eV} = 0.8\,\text{eV}$ ✓

(c) $E = hf = \dfrac{hc}{\lambda} = \dfrac{(6.63 \times 10^{-34}\,\text{J s} \times (3.00 \times 10^{8}\,\text{m s}^{-1})}{654 \times 10^{-9}\,\text{m}}$ ✓

$= 3.04 \times 10^{-19}\,\text{J} = \dfrac{3.04 \times 10^{-19}\,\text{J}}{1.6 \times 10^{-19}\,\text{J eV}^{-1}} = 1.9\,\text{eV}$ ✓

This is the transition from the −1.51 eV level to the −3.41 eV level. ✓

(e) The final mark is lost if the transition is the wrong way round, or if the minus signs are omitted.

Question 14

The diagrams show the cross-section of the wing of a model aeroplane in level flight and while ascending. The lines represent the motion of layers of air relative to the wing.

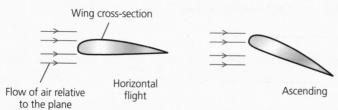

Wing cross-section

Flow of air relative to the plane

Horizontal flight

Ascending

(e) This question is about the flow of fluids relative to a body, and how the uplift on an aircraft wing is developed.

(a) In level flight the flow of air across the wings is streamlined (or laminar), but if the angle of ascent is too great, turbulence may occur and the aeroplane may stall.

Copy the diagram and complete the airflow lines above and below both wings. (2 marks)

(b) The model aeroplane has a mass of 250 g and the surface area of the underside of its wings is 0.10 m². Show that the pressure difference between the underside of the wing and upper surface is about 25 Pa when the model is flying horizontally. (3 marks)

Total: 5 marks

🅔 This is a 'show that' question and so the answer is given to three significant figures. Candidates often lose the first mark by using the mass of the aeroplane instead of its weight.

Student answer

(a) Laminar flow: continuous lines above and below the wing. ✓

Turbulent flow: some 'swirling' or break in continuity of the lines. ✓

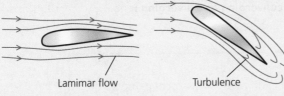

Lamimar flow   Turbulence

(b) $\Delta P = \dfrac{\text{upward force on wings (uplift)}}{\text{surface area of the underside of wings}}$

In level flight, the uplift = the weight of the aeroplane = $0.250 \, \text{kg} \times 9.8 \, \text{m s}^{-2}$ ✓

$\Delta P = \dfrac{0.250 \, \text{kg} \times 9.8 \, \text{m s}^{-2}}{0.10 \, \text{m}^2} = 24.5 \, \text{Pa}$ ✓

Question 15

The ray diagrams illustrate how an eye lens adjusts its shape to focus the rays from a distant object, also those from an object at the 'near point' 25 cm in front of the lens.

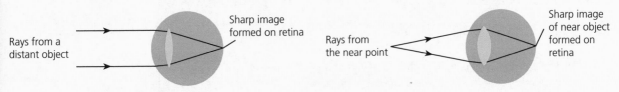

Rays from a distant object   Sharp image formed on retina   Rays from the near point   Sharp image of near object formed on retina

🅔 You should be aware that the 'accommodation' of the eye is achieved by making the lens fatter (more powerful) to focus on close objects, and by relaxing the muscles to allow the lens to return to its thinnest (least powerful) for distant objects.

The distance to the retina from this lens is 25 mm.

(a) Calculate the focal length of the lens in each situation, and determine the power of the lens in both cases   (4 marks)

A short-sighted person has an identical lens to that above, but the retina is 27 mm from the lens. At its thinnest, the lens focuses light from a distant object in front of the retina, so the image is blurred.

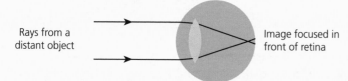

Rays from a distant object   Image focused in front of retina

A contact lens is placed close to the eye lens so that the combined lens focuses the light from the distant object onto the retina, forming a clear image

(b) (i) Calculate the power of the combined lens when it focuses on the distant object.   (1 mark)

(ii) Determine the power of the contact lens.   (2 marks)

(iii) State whether the contact lens is a converging or a diverging lens.   (1 mark)

Total: 8 marks

Student answer

(a) The parallel rays from the distant object will meet at the focal point, so the focal length will be 25 mm ✓ and $P = \dfrac{1}{f} = \dfrac{1}{0.025\ \text{m}} = 40$ dioptres ✓

$\dfrac{1}{u} + \dfrac{1}{v} = \dfrac{1}{f} = \dfrac{1}{250\ \text{mm}} + \dfrac{1}{25\ \text{mm}} \Rightarrow f = 23\ \text{mm}$ ✓ $P = \dfrac{1}{0.023\ \text{m}} = 44$ dioptres ✓

ⓔ When focused onto the distant object the focal length of the combination will equal the distance from the retina to the lens.

(b) (i) $f = 27\ \text{mm} \Rightarrow P = 37$ dioptres ✓

(ii) $P_{\text{combined}} = P_{\text{contact}} + P_{\text{lens}} \Rightarrow 37$ dioptres $= P_{\text{contact}} + 40$ dioptres ✓

(iii) $P_{\text{contact}} = -3$ dioptres ✓

ⓔ The minus sign is needed for the second mark

(iii) The contact lens is a diverging lens ✓ (because the power is negative indicating that the lens has a virtual focal point)

Question 16

(a) Draw a free-body force diagram of a sphere falling through a fluid at its terminal velocity. Label your diagram naming all the forces acting on the sphere.

(3 marks)

ℯ To gain all three marks, a candidate must draw all the forces along a vertical line through the centre of the sphere. The forces must meet at the centre, or be touching the surface of the sphere.

(b) With reference to the diagram, explain the meaning of 'terminal velocity'.

(2 marks)

ℯ To gain the second mark, some reference must be made to the free-body force diagram.

(c) A student carried out an investigation into the variation of the viscosity of syrup with its temperature. The terminal velocity of a 4.0 mm diameter ball bearing was found by timing the ball as it moved between two marks drawn 10.0 cm apart on the side of a test tube. This was repeated over a range of temperatures and the corresponding values of the viscosity of the syrup were calculated.

In her conclusion the student stated 'The accuracy of the experiment, particularly at higher temperatures, could be improved by using 2 mm diameter balls and a longer test tube.'

Comment on this conclusion, explaining if, and how, the terminal velocity measurements would be affected by these changes.

(3 marks)

Total: 8 marks

ℯ This investigation is based on **Core practical 4**. You are expected to know what factors affect the terminal velocity of a sphere falling through a fluid, and how the precision of the final results depends on how the measurements are made.

An A-grade candidate would answer this question by referring to the percentage uncertainties in the readings. Such a candidate may also realise that the terminal velocity depends on the square of the radius, and state that a sphere of half the radius will travel four times more slowly.

Student answer

(a) Weight, upthrust and viscous drag (resistive force or simply drag OK) correctly drawn and labelled. ✓✓✓

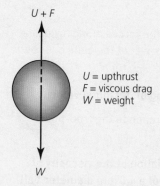

$U + F$

U = upthrust
F = viscous drag
W = weight

W

(b) At terminal velocity the resultant force on the sphere is zero. ✓

In the diagram, $W = U + F$; or the sum of the upward forces on the diagram equal the sum of the downward forces ✓

(c) If the terminal velocity is high, the time taken for the ball to fall between the marks will be small, leading to a bigger percentage uncertainty (error) in the value of the terminal velocity, and hence in the viscosity. ✓

ⓔ The uncertainty in the time will depend on the resolution of the scale Δt, but for a manually operated timing mechanism it is likely to be limited to ± 0.1 s because of reaction times. The percentage uncertainty $= \pm \dfrac{\Delta t}{t} \times 100\%$

The times will be shortest at higher temperatures because the viscosity will be lower and so the terminal velocity will be faster. ✓

The time can be made longer by using smaller balls or by increasing the length between the timing marks (either ✓).

Question 17

In microwave ovens, microwaves reflect from the metal walls. In basic models with no revolving tray or rotating reflectors, direct and reflected rays can interfere to produce hot and cold spots, leading to uneven cooking.

(a) Explain how these hot spots and cold spots are produced. (2 marks)

(b) A thin slice of cheese cooked in the oven was found to have small molten regions about 6 centimetres apart. Estimate the wavelength of the microwaves and show that the microwave frequency is approximately 2.5 GHz (2 marks)

The diagram of the microwave oven shows how two waves reach the point X having travelled along different paths. The reflected ray undergoes a phase change of π radians at Y.

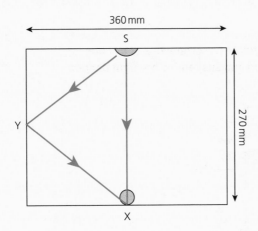

The source and X can be assumed to be at the midpoints of the upper and lower surfaces.

(c) Explain why there is a hot spot at X. (3 marks)

Total: 7 marks

ⓔ You should be aware that a phase change of π radians is equivalent to a path difference of half a wavelength.

Student answer

(a) If a microwave beam arriving directly from the source at a point combines with a beam that has been reflected from a wall, the waves will be superimposed on each other. If the microwaves are in phase at that point they will interfere constructively and create a high intensity wave at that point (a hot spot). ✓

If the waves are in antiphase, they will interfere destructively leading to a cold spot. ✓

ⓔ An explanation in terms of the production of antinodes and nodes in a standing wave will be credited here.

(b) The distance between adjacent antinodes in a standing wave = $\frac{\lambda}{2}$ = 6 cm → λ = 12 cm ✓

$$f = \frac{c}{\lambda} = \frac{3.0 \times 10^8 \text{ m s}^{-1}}{0.12 \text{ m}} = 2.5 \times 10^9 \text{ Hz} ✓$$

(c) By Pythagoras, SY = 225 mm so the path difference (SYX – SX) = (450 mm – 270 mm) = 180 mm ✓

For constructive interference to occur at X, the path difference (SYX – SX) would normally be equal to $N\lambda$ (N = an integer) but because of the phase change at Y the path difference needs to be $(N + \frac{1}{2})\lambda$. ✓

180 mm = 1.5 × 120 mm ✓ so there will be a hot spot at X

Question 18

A student performed an experiment to investigate the properties of a rubber band. She looped the band over a clamped rod and suspended a weight hanger from it as shown in the diagram.

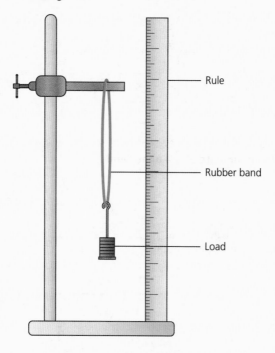

A metre rule was fixed vertically alongside the band. The student then added a series of masses to the hanger and measured the length of the band for each value of weight applied to the band.

She was advised by her teacher to take care to avoid parallax errors when measuring the length of the band.

(a) Explain what is meant by 'parallax' and describe how she could modify the arrangement in order to minimise the error. (3 marks)

ⓔ Simply stating that the eye is level with the bottom of the band and the scale is not enough here. Extra equipment or a modification of the set-up is needed.

The student's measurements are given in the table below:

W/N	$l/\times 10^{-2}\,\text{m}$	$\Delta l/\times 10^{-2}\,\text{m}$
0	12.0	0.0
2.0	14.0	2.0
4.0	17.5	5.5
6.0	29.5	17.5
8.0	34.0	22.0
10.0	36.0	24.0
12.0	37.0	25.0

(b) Use the student's results to plot a graph of W/N against $\Delta l/m$. (4 marks)

(c) Explain the shape of the graph in terms of the changes to the molecular structure of rubber during the extension. (3 marks)

(d) Describe a method of estimating the work done by the weights in stretching the band from the graph. (2 marks)

ⓔ Note that the question does not ask you to determine a value for the work done.

(e) If the weights were removed one at a time and the extensions were found for each value, sketch the position of the unloading curve on your graph. (1 mark)

(f) Use your answers to parts (d) and (e) to explain why the band will become warmer if it is repeatedly stretched and released over a period of time. (2 marks)

Total: 15 marks

Student answer

(a) If there is a gap between the bottom of the band and the scale, the position of this point will appear to be at different positions on the scale when viewed from different angles. ✓

To avoid parallax errors the eye must be level with the bottom of the band so that the line of view is perpendicular to the scale. ✓

To ensure that this is the case, a set-square should be held against the scale or a pointer fixed to the bottom of the band so that it touches the scale. ✓

(b) Graph (e.g. drawn on a 10 cm × 10 cm grid); axes labelled W/N and $\Delta l/cm$ ✓

Suitable scale e.g. 1.0 N / division and 2.5 cm / division ✓

All 6 points accurately plotted ✓

Smooth curve drawn through the plots (as in the top line on Figure 15, p.19) ✓

(c) The rubber is made from long-chain molecules (polymers) that are initially loosely tangled together. ✓

When the weights are added, the chains untangle leading to a large extension for each added weight, so the gradient is small. ✓

When the molecules are untangled, a much larger force is needed to stretch the bonds of the molecules, so the gradient of the graph increases markedly. ✓

ⓔ It is important that the second and third points refer to the shape of the graph as well as the changes in molecular structure.

(d) The work done is found by estimating the area under the curve and then multiplying by the amount of work represented by each square. ✓

The area can be found by determining the area of each small square and counting the number of squares, or by approximating the shape to a number of rectangles and triangles, and adding the values of their areas. ✓
[any two points]

(e) The unloading line is below the loading line (see lower line on Figure 15 p. 19) ✓

(f) The area between the loading and unloading lines represents the difference in the work done on the band during stretching and that done by the band in contracting (hysteresis) ✓

This work is transferred as internal energy in the rubber so that its temperature will increase. ✓

Question 19

The beam from a red laser pen is directed through a diffraction grating. The diffraction pattern seen on a screen behind the grating is shown next to a scale fixed to the screen.

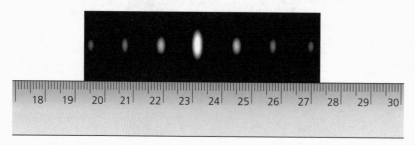

(a) To determine the wavelength of the light it is necessary to measure the separation, x, between the central maximum and one of the other maxima. Use the scale to measure a value of x, such that your measurement has the least percentage uncertainty.

(2 marks)

e This is an exercise in measurement technique. You are required to measure the distance that will lead to a value of wavelength that has the least uncertainty.

(b) Show how you calculated the uncertainty.

(1 mark)

(c) The diffraction grating has a groove density of 50 lines mm⁻¹, and the distance of the grating from the screen is 39 cm.

   (i) Calculate the wavelength of the laser light.

(3 marks)

   (ii) Describe the change in the pattern if a green laser pen is used.

(1 mark)

(d) When fast-moving electrons are fired through a thin wafer of carbon, circular diffraction patterns can be observed on a fluorescent screen beyond the wafer.

(i) What can be deduced from this observation about the nature of the electrons?

(1 mark)

(ii) Calculate the wavelength of an electron travelling at a speed of $6.2 \times 10^6 \, \text{m s}^{-1}$

(3 marks)

ⓔ You will need to look up values of the mass of the electron and Planck's constant from the data sheet, and also be aware that momentum = mass × velocity.

(iii) Explain why the radii of the circular rings decrease when the speed of the electrons is increased.

(2 marks)

Total: 13 marks

Student answer

(a) Distance from the 3rd order maximum on the left to the 3rd order on the right; because this is the largest possible measurement and so will have the smallest percentage uncertainty ✓

$2x = 75 \, \text{mm} \Rightarrow x = 38 \, \text{mm}$ ✓. Any correct value of distance from the central maximum to another maximum will gain one mark.

(b) Percentage uncertainty = $\dfrac{\pm 1 \, \text{mm}}{75 \, \text{mm}} \times 100\% = \pm 1.3\%$ ✓

ⓔ The uncertainty in measurements from a mm scale is usually ± 0.5 mm, but the images are a bit blurred so ± 1 mm or ±0.5 mm are acceptable.

(c) (i) $3\lambda = \dfrac{1}{50 \times 10^3 \, \text{m}^{-1}} \sin \theta$ ✓; $\tan\theta = \dfrac{3.75 \, \text{cm}}{39 \, \text{cm}} \rightarrow \sin\theta = 0.0957$ ✓

$\lambda = \dfrac{2.0 \times 10^{-5} \, \text{m} \times 0.0957}{3} = 6.4 \times 10^{-7} \, \text{m} = 640 \, \text{nm}$ ✓

(ii) If green light is used, the wavelength is shorter than the red light, so the spaces between the maxima will be reduced. ✓

(d) (i) Diffraction is a wave property so the electron must behave as a wave as well as a particle. ✓

(ii) momentum = mass × velocity = $9.1 \times 10^{-31} \, \text{kg} \times 6.2 \times 10^6 \, \text{m s}^{-1}$ ✓

$\lambda = \dfrac{h}{p} = \dfrac{6.63 \times 10^{-34} \, \text{J s}}{5.64 \times 10^{-24} \, \text{N s}}$ ✓ $= 1.2 \times 10^{-10} \, \text{m}$ ✓

(iii) When the speed of the electron increases, its momentum increases and so its wavelength will become smaller. ✓

Smaller wavelengths show less diffraction and the rings will become closer together. ✓

Question 20

The fundamental frequency of the sound emitted by a stretched wire is given by the equation:

$$f = \frac{1}{2l}\sqrt{\frac{T}{\mu}}$$

where T is the tension in the wire, l is the length of the wire and μ is its mass per unit length.

(a) Write a plan for experiments to check the validity of the equation for a particular piece of wire. Your plan should include:

(i) A list of all materials to be used.

(ii) A fully labelled diagram of the apparatus.

(iii) The instruments to be used for each reading.

(iv) Comments on how the experiment will be performed safely.

(v) A discussion on how the data collected will be used to verify the relationship

(8 marks)

ⓔ This question is based on **Core practical 7**. You are expected to be familiar with the experiment and the details of measurements, precautions, safety and how the data is analysed.

(b) Explain how you would use the results of one experiment to determine the value of μ for the wire.

(2 marks)

Total: 10 marks

Total for paper: 90 marks

ⓔ There are several methods of producing a fundamental standing wave in a wire. Usually one end of the stretched wire is attached to an oscillator with the tension provided by weights on a hanger on the other end using a wire looped over a pulley wheel. A second method has the wire passing between the poles of a magnet with the ends of the wire connected to a signal generator.

Student answer

(a) (i) Length of wire, signal generator, pulley wheel, oscillator or magnet, hanger plus weights, metre rule, connecting leads, balance. ✓ (The frequency output from the signal generator is usually checked using a strobe lamp or a cathode ray oscilloscope, and masses are checked using a top pan balance.)

(ii) Fully labelled diagram. ✓

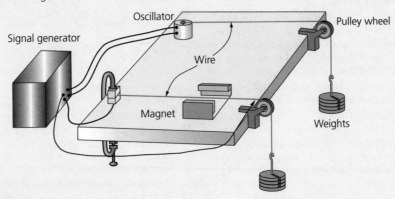

(iii) The length of the wire is measured using a metre rule. ✓

The masses of the weights are checked using a balance. ✓

The frequency of the supply is measured from the scale of the signal generator or using a strobe lamp or CRO. ✓

The mass of a measured length of wire is found using a balance. ✓

(iv) Safety glasses must be worn; and a tray of sand/foam rubber should be placed under the weights to prevent them falling onto feet, or other sensible precaution (any one). ✓

(v) Either one of:

- Keeping l and μ constant, take values of f for a range of values of T. ✓ Plot a graph of f against $\sqrt{T}$, or f^2 against T. ✓
- Keeping T and μ constant, take values of f for a range of values of l. ✓ Plot a graph of f against $\frac{1}{l}$. ✓

Maximum 8 marks

(b) Find μ from the gradient of the graph ✓

When l is fixed, the gradient of the graph will be $\dfrac{1}{2l\sqrt{\mu}}$ or $\dfrac{1}{4l^2\mu}$ ✓

or

When T is fixed, the gradient will be $\dfrac{1}{2}\sqrt{\dfrac{T}{\mu}}$ ✓

Knowledge check answers

Knowledge check answers

1. $P = h\rho g$

   Mercury: $P = 0.760\,\text{m} \times 13\,600\,\text{kg m}^{-3} \times 9.8\,\text{m s}^{-2} = 1.0 \times 10^5\,\text{Pa}$

   Water: $P = 10\,\text{m} \times 1000\,\text{kg m}^{-3} \times 9.8\,\text{m s}^{-2} = 1.0 \times 10^5\,\text{Pa}$

2. The upthrust on it must equal its weight; it must displace its own weight of fluid.

3. As fluid molecules move relative to each other; the motion is opposed by the attractive (cohesive) forces between them. They also 'stick' to the walls of the tubes because of the adhesive forces between the solid molecules and the fluid molecules.

4. Factors: the densities of the fluid and the sphere; the radius of the sphere; the viscosity of the fluid
   Affects most: the radius as $v \propto r^2$

5. The initial length of the wire (metre rule); the load applied (standard masses – checked using a balance); the extension of the wire (marker along a mm scale); the diameter (*not* the radius or area) of the wire (micrometer)

6. Stress–strain graphs are independent of the dimensions of the wire; the Young modulus can be found directly from the gradient; the strength (UTS) from the maximum breaking stress.

7. As the rubber is deformed and reformed; it goes through a stress–strain cycle; there is a large area under the graph; suggesting that the tyres absorb a large amount of energy per unit volume during use.

8. a microwaves
   b ultraviolet
   c radio

9. a visible
   b infrared
   c X-rays or gamma rays

10. Inverse square law; because the distance is increased 4 times; the intensity will decrease 16 times; to $0.5\,\text{W m}^{-2}$

11. Coherent sources; of similar amplitude

12. Boxes labelled microphone, amplifier, (electronic) mixer and headphones; connected in order by arrows.

13. $\lambda = \dfrac{v}{f} = \dfrac{340\ \text{m s}^{-1}}{256\ \text{s}^{-1}} = 1.33\ \text{m}$

14. $\mu = \dfrac{5.0 \times 10^{-3}\text{kg}}{0.80\ \text{m}} = 6.25 \times 10^{-3}\text{kg m}^{-1}$

    $v = \sqrt{\dfrac{T}{\mu}} = \sqrt{\dfrac{32\ \text{N}}{6.25 \times 10^{-3}\text{kg m}^{-1}}} = 72\ \text{m s}^{-1}$

15. a $n = \dfrac{3.00 \times 10^8 \text{m s}^{-1}}{2.10 \times 10^8 \text{m s}^{-1}} = 1.43$

    b $1.43 = \dfrac{\sin 30°}{\sin r} \rightarrow r = 20°$

16. $\sin C = \dfrac{n_{\text{cladding}}}{n_{\text{core}}} = \dfrac{1.48}{1.56} \rightarrow C = 72°$

17. $P_1 = \dfrac{1}{0.20\ \text{m}} = +5.0$ dioptres;

    $P_2 = \dfrac{1}{0.25\ \text{m}} = +4.0$ dioptres

    $P = +5.0$ dioptres $+ 4.0$ dioptres $= +9.0$ dioptres

18. Real images are formed by the actual intersection of rays of light; and so can be formed on a screen. Virtual images are at positions where the rays of light appear to come from; but do not actually meet; they cannot be formed on a screen.

19. Sound waves are longitudinal; these waves cannot be polarised; because the oscillations of the particles are always parallel to the direction of the wave motion

20. The waves the mobile phone receives have much shorter wavelengths than radio waves; there will be less diffraction (spreading of the waves around the obstruction)

21. If a continuous wave were used; all the reflections from the surfaces would be received all the time; making it impossible to distinguish specific echoes

22. a $E = hf = 6.63 \times 10^{-34}\,\text{Js} \times 7.2 \times 10^{15}\,\text{s}^{-1} = 4.8 \times 10^{-18}\,\text{J}$

    b $E = \dfrac{hc}{\lambda} = \dfrac{6.63 \times 10^{-34}\,\text{Js} \times 3.00 \times 10^8\ \text{m s}^{-1}}{6.3 \times 10^{-11}\ \text{m}} = 3.2 \times 10^{-15}$

23. $\phi = hf - E_{k\,\text{max}} = (9.3 \times 10^{-19}\,\text{J}) - (1.3 \times 10^{-19}\,\text{J})$

    $= 8.0 \times 10^{-19}\,\text{J}$

    $= \dfrac{8.0 \times 10^{-19}\ \text{J}}{1.6 \times 10^{-19}\ \text{J eV}^{-1}} = 5.0\ \text{eV}$

24. a The energy level of an electron that can exist in a stable orbit
    b The displacement of an electron from one permitted orbital to another
    c The minimum energy required to remove an electron from an atom; from its ground state

Note: page numbers in **bold** indicate key term definitions.

A

absolute refractive index **38**
A-level test paper 77–93
amplitude 21, 27–28, 32, 33–34
antinodes, standing waves 32, 35, 36
antiphase, waves 26, 27, 28
Archimedes' principle **8**
AS test paper 61–76
atomic spectra 56–57

B

Bohr, Niels 51, 56
buoyancy force (upthrust) 8

C

coherent wave sources 29
compression, solids 14–19
constructive interference 30, 31, 66, 87
constructive superposition 27–28, 30
converging lenses 41, 42
critical angle **40**, 41

D

de Broglie, Louis 52, 58
density **6**
destructive interference 32, 66
destructive superposition 28, 30
diffraction **46–47**
diffraction gratings 47–48
dioptre 43, 44
displacement–distance/time graphs 20–21, 25–27
diverging lenses 41, 42
drag 10–11, 13

E

eddy currents 9
Edexcel exams 59
Einstein, Albert 51, 53
Einstein's photoelectric equation 53, 54

elastic limit 15
elastic strain energy 18–19
electromagnetic waves 23–24
electronvolt (eV) **52**
emission spectrum **56**
extension, solids 14–19

F

fibre optics 40–41
fluids 6–14
    density of 6–7
    flow of 9
    pressure in 7
    Stoke's law 10–12
    upthrust 8
    viscosity 9–13
focal length 42, 43
focal point **42**
force–extension graphs 14–15
free-body force diagrams 10–11
frequency 21
    standing waves 34–37
    threshold, photoelectrons 53–55
fundamental frequency 36, 92

G

gases *see* fluids
ground state 56

H

harmonics 36
Hooke's law **14**, 19
Hooke's law region 14, 15, 16
Huygens, Christiaan 51
Huygens' construction 29
hysteresis in rubber 19–20

I

images, real and virtual 42, 43
infrared waves 24
intensity **24–25**
interference 29–31
interference patterns 30–31, 32
inverse square law 25
ionisation energy/level 56, 57

K

kinetic energy, photoelectrons 53–55, 72

L

laminar flow 9
lens equation 43
lenses 41–44
    power of 43–44
    ray diagrams 42–43
light
    intensity of 24–25
    particle nature of 51–52
    photoelectric effect 53–55
    polarised 44–46
    ray diagrams 42–43
    reflection 49
    refraction 38–39
limit of proportionality 15
liquid crystal displays (LCDs) 46
liquids *see* fluids
longitudinal waves **22**
    sound waves 33–37

M

maximum kinetic energy of electrons 54, 55
measurement uncertainty (error) 85–86, 90–91
microwaves 24, 45, 86–87

N

Newton, Sir Isaac 51, 52, 58
nodes, standing waves 32, 35, 36
noise control 31

O

optical centre 43
orbitals, electrons 56
overtones, standing waves 36

P

particle nature of light 51–52, 58
path difference 30–31, 32, 66, 87
percentage uncertainty (error) 85–86, 90–91

Index

period (of a wave) 21

permitted orbitals 56

phase and phase difference, waves 25–27

photoelectric effect 53–55

photon energy 53

Planck, Max 51

Planck's constant 52, 53, 58

Planck's equation 52, 53

plane polarisation 44–46, 68–69

plane polarised waves **44**

plastic flow 15

polarisation of light 44–46

Polaroid® 44

potential difference 52, 54

power (of lenses) **43–44**

practical work

    frequency of vibrating string 35

    speed of sound 23

    viscosity of a liquid 11

    wavelength of light 48

    Young modulus 16

pressure in fluids 7

principal axis 43

progressive waves 32–33

pulse–echo techniques 49–50

Q

quantised orbits, electrons 56

R

radiation 23–24

radiation flux 24–25

radio waves 24

ray diagrams 42–43, 83–84

real image 42, 43

real is positive sign convention 43

reflection

    light 49

    microwaves 86–87

    sound 49–50

total internal 39–41

    ultrasound 50, 69–71

refraction 37–38

    lenses 41

    of light 38–39

refractive index **38**, 39, 40–41

rubber, hysteresis in 19–20

Rutherford, Ernest 51, 56

S

Snell's law **38**, 40

solids

    properties of 19–22

    tensile behaviour of 15–18

sound waves 22, 33–34, 36–37

spectra, atomic 56–57

spectrum, electromagnetic 23–24

speed of sound 23, 31, 34, 37

spring constant 14, 78

standing waves 31–34

    in air 36–37

    in strings 34–36

stationary waves 32–33

Stokes' law **10**

stopping potential 54

strain **15**

streamlined flow 9

stress **15**

stress–strain graphs 16–18

strings, standing waves in 34–36

superposition 27–28

T

temperature and speed of sound in air 37

tensile forces 14–18

terminal velocity 11, 12, 13, 85–86

threshold frequency 53–54

total internal refraction 39–41

transverse waves **22–23**

turbulent flow 9

U

ultimate tensile stress (UTS) 16, 17–18

ultrasound 50, 69–71

ultraviolet 24, 53–54

uncertainty in measurement 85–86, 90–91

upthrust 8–9

V

virtual image 42, 43

viscometer 10, 66–67

viscosity 9–13, 85–86

visible light 24, 53–54

voltage 52, 54

vortices 9

W

wave equation 21

wavefronts **28–29**

    diffraction 46–47, 64

wavelength 21, 23–24

wave–particle duality 58

waves 20–58

    electromagnetic 23–24, 45

    interference 29–31

    longitudinal 22

    plane polarised 44–46

    standing 31–37

    superposition 27–28

    terminology 20–22

    transverse 22–23

wave speed 21, 27, 38

wind instruments 36–37

work function **53**

X

X-ray diffraction patterns 48

X-rays 23

Y

yield point 15

Young modulus **15–16**